Lakhdar Meziani
Liaqat Ali Khan

Vector Measures, Integration and Integral Representation Theory

Lakhdar Meziani
Liaqat Ali Khan

Vector Measures, Integration and Integral Representation Theory

Noor Publishing

Imprint
Any brand names and product names mentioned in this book are subject to trademark, brand or patent protection and are trademarks or registered trademarks of their respective holders. The use of brand names, product names, common names, trade names, product descriptions etc. even without a particular marking in this work is in no way to be construed to mean that such names may be regarded as unrestricted in respect of trademark and brand protection legislation and could thus be used by anyone.

Cover image: www.ingimage.com

Publisher:
Noor Publishing
is a trademark of
International Book Market Service Ltd., member of OmniScriptum Publishing Group
17 Meldrum Street, Beau Bassin 71504, Mauritius

Printed at: see last page
ISBN: 978-620-2-34954-3

Vector Measures, Integration and Integral Representation Theory

Lakhdar Meziani[a] **and Liaqat Ali Khan**[b]

[a]Department of Mathematics, Faculty of Science, University of Batna, Algeria
[b]Department of Mathematics, Quaid-i-Azam University, Islamabad, Pakistan

ABSTRACT The objective of this monograph is intended to present a unified survey of the theory of integral representation of bounded operators in topological spaces. Beside for those researchers in this area, it might be of interest for the graduate students who are familiar with abstract measure and integration. The material of the theory starts with the famous representation theorem of Riesz which asserts that bounded linear functionals on spaces of continuous real-valued functions on compact space have an integral form with respect to Lebesgue measure.

In this work, we will be concerned with some prominent generalizations of this Theorem, frequently used in the literature. We will essentially use the settings of Banach spaces and locally convex spaces

Contents

Preface

The aim of this monograph is intended to present a unified, mostly self-contained, survey of the theory of integral representation of bounded operators in topological spaces. It is written for the graduate students who are familiar with abstract measure and integration, and also with the main topics of Banach spaces and topological vector spaces.

The starting material of the theory seems to be the representation theorem of Riesz which asserts that if S is a compact Hausdorff space, and $C(S)$ is the space of continuous real-valued functions on S, with the uniform norm, then for each bounded linear functional $T : C(S) \to \mathbb{R}$, there is a unique bounded regular measure μ on the Borel σ-field $\mathcal{B}(S)$ of S such that:

$$Tf = \int_S f \, d\mu \ \text{ for all } \ f \in C(S)$$

where the integral is a Lebesgue one, and $\|T\| = |\mu|$, the variation of μ.

This statement of the theorem, due to **Kakutani [Kak41]**, is one of the versions we have at hand nowadays.

A lot of work has been done on various extensions of this theorem and it is still the object of many investigations. The objective of this work is a presentation of some prominent generalizations of **Riesz Theorem**, frequently used in the literature. We will essentially be concerned by the following settings:

In the first part, mainly covered by chapters 1 to 3, we consider Banach spaces X, Y , and form the Banach space $C(S, X)$ of all continuous functions $f : S \to X$, from the compact space S into X, equipped with the uniform norm, put $C(S, X) = C(S)$ if $X = \mathbb{R}$. The main problem we will be concerned with is that of representing linear bounded operators $T : C(S, X) \to Y$ in an integral form:

$$Tf = \int_S f \, d\mu, \ \ f \in C(S, X),$$

for some integration process with respect to a measure μ on the Borel σ-field $\mathcal{B}(S)$ of S.

In the second part covered by chapters 4 to 6, the objective is to go beyond the Banach space setting, to a topological vector space (TVS) context. In this part, X will be a locally convex space with dual X^* and S a locally compact space. We denote by $C_0 (S, X)$ the function space of all continuous functions $f : S \to X$, vanishing outside a compact set of S, put $C_0 (S, X) = C_0 (S)$ if $X = \mathbb{R}$. We are interested in representing linear bounded operators $T : C_0\,(S, X) \to X$, by means of weak integrals, specifically Pettis integrals against scalar measure μ on S. Each representation theorem considered here has its own integration process which goes through duality of function spaces, given in details in chapter 2.

Introduction

One of the most famous theorems in integration theory is the representation theorem of Riesz, which asserts that if S is a compact Hausdorff space, and $C(S)$ is the space of continuous real-valued functions on S, with the uniform norm, then for each bounded linear functional $T : C(S) \to \mathbb{R}$, there is a unique bounded regular measure μ on the Borel $\sigma-$field $\mathcal{B}(S)$ of S such that:

$$Tf = \int_S f \, d\mu \quad \text{for all} \ \ f \in C(S)$$

where the integral is a Lebesgue one, and $\|T\| = |\mu|$, the variation of μ.

This statement of the theorem, due to Kakutani [**KaKu41**], is the frequently used version we have at hand nowadays.

In this monograph we will be concerned with some major extensions of this theorem in general settings. A lot of work has been done in this domain (see [**BDS55,DU77, Din67, DS58, Goo70**] and the references. therein.), and it is still the object of many investigations [**BoD01, Mez08, Khu09**]. The objective of this work is a presentation of some prominent generalizations of Riesz Theorem, frequently used in the literature. We will manly consider the following settings:

In the first part, mainly covered by chapters 1 to 3, we consider Banach spaces X, Y , and form the Banach space $C(S, X)$ of all continuous functions $f : S \to X$, from the compact space S into X, equipped with the uniform norm, put $C(S, X) = C(S)$ if $X = \mathbb{R}$.

The main problem we will be concerned with is that of representing linear bounded operators $T : C(S, X) \to Y$ in an integral form:

$$Tf = \int_S f \, d\mu \ \ f \in C(S, X),$$

for some integration process with respect to a measure μ on the Borel $\sigma-$field $\mathcal{B}(S)$ of S. Each representation theorem considered here has its own integration process. In chapter 1 we give material pertaining to vector integration which will be used. Chapter 2 is concerned with duality in some function spaces equipped with appropriate topologies. In chapter 3 we give three major representation theorems. We start with the famous theorem of Bartle-Dunford Schwartz [**BDS55**], representing general $X-$valued operators on $C(S)$. In this representation, the class of weakly compact operators has the important property of being represented by integrals against vector measures with values in the Banach space X.

Next we turn to Dinculeanu-Singer theorem [**DU77**], for general bounded operators $T : C(S, X) \to Y$. The theorem goes through the structure of the topological dual $C^*(S, X)$ of the function space $C(S, X)$, given in [**Sing57**], and generalized in [**Mez09c**], see chapter 2 for details The aim of the third representation theorem is the construction of a class of operators from $C(S, X)$ into X, characterized by their Bochner integral, given in [**Mez02**].

In the second part, covered by chapters **4** to **6**, the objective is to go beyond the Banach space setting, to a topological vector space (TVS) context. In this part, X will be a locally convex space with dual X^* and S a locally compact space. We denote by $C_0(S, X)$ the function space of all continuous functions $f : S \to X$, vanishing outside a compact set of S, put $C_0(S, X) = C_0(S)$ if $X = \mathbb{R}$. We are interested in representing linear bounded operators $T : C_0(S, X) \to X$, by means of weak integrals against scalar measure μ on S. First, in section **4.1**, we start with an operator $T : C_0(S, X) \to X$ and give conditions under which T can be written as a Pettis integral with respect to a scalar measure μ. Second, we consider the converse, that is, given a measure μ of bounded variation on $\mathcal{S}$, we seek for an operator $T : C_0(S, X) \to X$, which will have a Pettis integral form with respect to μ. This is a more delicate problem which needs additional assumptions on the space X. Solutions to this problem [**Mez08**], under various conditions on the dual X^*, are given in section **4.2**.

In Chapter. **5**, we consider two extensions of the representations given in sections **3.1** and **3.2**.

The first one, due to D.R. Lewis [**Lew70**], develops an appropriate integration process with respect to measures into a locally convex Hausdorff space X, and gives integral representation for a weakly compact operator $T : C(S) \longrightarrow X$, with respect to an $X-$valued measure in the case of a compact Hausdorff S. This extends the representation of Bartle-Dunford-Schwartz given in Theorem **3.1.5** for a Banach space X.

The second representation is similar to that of Dinculeanu-Singer given in **Theorem 3.2.1**. It is due to A. Katsaras and D. B. Liu [**KL 75**], who considered an entirely different setting from that used by Dinculeanu-Singer. They have been able to get a representation for weakly compact operators, with respect to an operator valued measure for some special function spaces (see section **5.3** below for details).

Chapter 6 presents a theory of integral representation whose settings goes beyond the locally convex TVS, namely to general TVS and especially to not locally convex TVS. As usual, the theory starts with the description of the integration process which will give the integral representation in an appropriate context. This theory has been developed by A.H. Shuchat [**Shu72**]. It includes generalizations of the Fichtenholz-Kantorovick-Hilderbrandt and Riesz representation theorems to this setting, using strong integrals. The description of the integration process is given in section **6.1**. The structure of the dual of the function space $C(S, X)$ follows in section **6.2**. Integral representation of operators on the space $C(S, X)$ is given in section **6.3**.

Acknowledgement

The authors wish to acknowledge with thanks the Deanship of Scientific Research, King Abdulaziz University, Jeddah, for their technical and financial support.

1

Vector Integration

1.1 Vector Measures and Vector Measurable Functions

In what follows S will be an abstract set, $\mathcal{A} = \mathcal{A}(S)$ a $\sigma-$algebra of subsets of S, X a Banach space and X^* be the topological dual of X. Most of the material in this chapter can be found in standard books (e.g., [**DS58**]).

Definition 1.1.1. (i) A set function, $\mu : \mathcal{A} \to X$ is called $\sigma-$**additive** if for every pairwise disjoint sequence of sets $\{E_n\}$ in $\mathcal{A}$, the series $\sum_n \mu(E_n)$ is unconditionally convergent in X and we have

$$\mu \left(\bigcup_n E_n \right) = \sum_n \mu(E_n).$$

If, in addition, $\mu(\varnothing) = 0$, then μ is called a **vector measure**.

(ii) A set function $\mu : \mathcal{A} \to X$ is called **weakly** $\sigma-$**additive**. if for every pairwise disjoint sequence of sets $\{E_n\}$ in $\mathcal{A}$ we have

$$x^* \mu \left(\bigcup_n E_n \right) = \sum_n x^* \mu(E_n)$$

for each $x^* \in X^*$, in other words the real set function $x^* \mu$ is a real measure.

Remark 1.1.2. It is clear that a vector measure is weakly $\sigma-$additive. The converse is also true (see **Theorem 1.1.5** below).

Definition 1.1.3. The **semi-variation** of the vector measure μ is defined by the set function:

$$\|\mu\| (E) = \sup \left\| \sum_{i=1}^n \varepsilon_i \, \mu(E_i) \right\|, \quad E \in \mathcal{A},$$

the supremum being taken over all finite partitions $\{E_i\}$ of E in $\mathcal{A}$, and all finite systems of scalars $\{\varepsilon_i\}$ with $|\varepsilon_i| \leq 1$.

The semi-variation so defined is needed for some estimations in the integration process which will be used.

Definition 1.1.4. If $\mu : \mathcal{A} \to X$ is a vector measure, the **variation** of μ is defined by the positive set function $|\mu| (\bullet)$ given by

$$|\mu| (E) = \sup \sum_i \|\mu(E_i)\|, \quad E \in \mathcal{A}.$$

the supremum being taken over all finite partitions $\{E_i\}$ of E in $\mathcal{A}$.The notation $v (\mu, E)$ is also used for $|\mu| (E)$ by some authors [**DS58**].

We say that μ is of bounded variation if $|\mu|(S) < \infty$. Moreover, if μ is scalar valued, then we have (see Section **A.6**, or [**DS58, Lemma III.1.5**]):

$$|\mu|(E) \leq 4 \sup\{|\mu(F)|, F \in \mathcal{A}, \ F \subseteq E\}. \tag{\#1.1}$$

Theorem 1.1.5. *Let* $\mu : \mathcal{A} \to X$ *be a weakly* $\sigma-$*additive set function. Then:*
(a) μ *is a vector measure.*
(b) *Moreover we have:*

$$\|\mu\|(E) \leq 4 \sup\{\|\mu(F)\|, F \in \mathcal{A}, \ F \subseteq E\}, E \in \mathcal{A}.$$

Proof: (a) This part is due to Pettis (see [**BDS55**]).

If $\{E_n\}$ is a sequence of sets in $\mathcal{A}$ let $\mathcal{A}_0$ be the algebra generated by $\{E_n\}$ and let $\mathcal{A}_1$ be the $\sigma-$algebra generated by $\mathcal{A}_0$. Then the closed subspace X_1 of X generated by the family $\{\mu(E), E \in \mathcal{A}_0\}$, is separable. We claim that $\mu(F) \in X_1$, if $F \in \mathcal{A}_1$. If this is not the case for some $F \in \mathcal{A}_1$, then by the Hahn-Banach Theorem (**Section A.2**), there is $x^* \in X^*$ such that

$$x^*\mu(F) \neq 0 \text{ and } x^*\mu(E) = 0 \text{ for all } E \in \mathcal{A}_0.$$

This contradicts the uniqueness of the extension of the scalar measure $x^*\mu$ from the algebra $\mathcal{A}_0$ to the $\sigma-$algebra $\mathcal{A}_1$.

Now assume that μ is not $\sigma-$additive, then for some $\varepsilon > 0$ there exists a decreasing sequence $\{E_n\}$ whose intersection is empty such that $|\mu(E_n)| > \varepsilon$, for all n. By the **Hahn-Banach Theorem** (**Section A.2**), there is a sequence $x_n^* \in X^*$ with $|x_n^*| = 1$ and

$$x_n^*\mu(E_n) = |\mu(E_n)| > \varepsilon \text{ for all } n.$$

Consider $\mathcal{A}_1$ and X_1 as defined above and let (x_k) be a sequence dense in X_1. Using a **Cantor diagonal procedure** [**Rud87, p. 26**] (also see **Section A.1**), there is a subsequence $y_m^* = x_{n_m}^*$ such that $\lim_m y_m^*(x_k)$ exists for all k. Since $|y_m^*| = 1$, for all m. It follows from uniform boundedness principle (**Section A.2**) that $\lim_m y_m^*(x)$ for every $x \in X_1$ and in particular $\lim_m y_m^*(\mu(E))$ exists for each $E \in \mathcal{A}_1$. By the **Nikodym Theorem** (**Section A.6**), the set $\{y_m^*\mu\}$ of scalar measures is uniformly $\sigma-$additive on $\mathcal{A}_1$, which contradicts the assumption that $y_m^*\mu(E_n) > \varepsilon$ for all m.

(b) Next, for any $x^* \in X^*$, $x^*\mu$ is a finite scalar measure, so that ,

$$\sup\{|x^*\mu(F)|, F \in \mathcal{A}, \ F \subseteq E\} < \infty \text{ for each } x^* \in X^*.$$

By the Uniform Boundedness theorem (**Section A.2**), we deduce that

$$\sup\{\|\mu(F)\|, F \in \mathcal{A}, \ F \subseteq E\} < \infty.$$

On the other hand we have from **Definition 1.1.3**,

$$
\|\mu\|(E) \;=\; \sup\left|\sum_{i=1}^{n}\varepsilon_i\,\mu(E_i)\right| \;=\; \sup_{|x^*|\le 1}\sup\left|\sum_{i=1}^{n}\varepsilon_i\,x^*\mu(E_i)\right|
$$

$$
\le \;\sup_{|x^*|\le 1}\sup\sum_{i=1}^{n}\varepsilon_i\,|x^*\mu(E_i)| \;\le\; \sup_{|x^*|\le 1}|x^*\mu|(E)
$$

$$
\le \;4\sup_{|x^*|\le 1}\sup\{|x^*\mu(F)|,\,F\in\mathcal{A},\ F\subseteq E\}
$$

$$
= \;4\sup\{\|\mu(F)\|,\,F\in\mathcal{A},\ F\subseteq E\}.
$$

where the 2nd inequality comes from $(\#\mathbf{1.1})$ above. $\square$

Definition 1.1.6. (a) A $\mu-$**null set** is a subset of a set $E\in\mathcal{A}$ such that, $\|\mu\|(E)=0$. A property on S is said to be valid $\mu-$**almost everywhere** if it is valid on the complement of a $\mu-$null set. From now on, we will assume that $\mathcal{A}$ contains the $\mu-$null sets of S, otherwise, the Lebesgue completion process can be applied to $\mathcal{A},\|\mu\|$.

(b) We say that a function $f:S\to\mathbb{R}$ is **measurable** if for every Borel set B of $\mathbb{R}$, $f^{-1}(B)\in\mathcal{A}$.

(c) A **simple measurable function** of S into $\mathbb{R}$ is a finite linear combination of characteristic functions of pairwise disjoint sets in $\mathcal{A}$.

As is well known, we have:

Proposition 1.1.7. *A function $f:S\to\mathbb{R}$ is measurable if and only if it is the limit $\mu-$almost everywhere of a sequence of simple measurable functions. Also, the limit $\mu-$almost everywhere of a sequence of measurable functions, is measurable.*

Definition 1.1.8. Let $f:S\to\mathbb{R}$ be a real measurable function. We define the $\mu-$**essential supremum** of f on a set $E\in\mathcal{A}$ by:

$$
\mu - ess\,\sup_{s\in E}|f(s)| = \inf\{A:\|\mu\|(E\cap\{s:|f(s)|>A\}=0)\}
$$

if

$$
\mu - ess\,\sup_{s\in E}|f(s)| < \infty,
$$

we say that f is $\mu-$**essentially bounded** on the set E and $\mu-$essentially bounded if $E=S$.

In **Section 1.3** below, we need to integrate vector valued functions, so we have to make precise the concept of measurability for such functions.

Definition 1.1.9. (i) .An **elementary measurable function** $f:S\to X$ is a function of the form

$$
f(\bullet) = \sum_i \chi_{A_i}(\bullet)\,x_i,\ \text{or in short}\ f = \sum_i \chi_{A_i}\otimes x_i,
$$

where $\{A_i\}$ is a countable partition of S in $\mathcal{A}$ and $\{x_i\}$ a sequence of vectors in X. We denote by $\mathcal{E}(S, X)$ the set of all elementary measurable functions $f : S \to X$.

(ii).A function $f : S \to X$ is said to be **strongly measurable** if there is a sequence of elementary measurable functions f_n converging $\mu-$almost everywhere to f. Let $F_{sm}(S, X)$ be the set of all strongly measurable functions $f : S \to X$.

(iii).A function $f : S \to X$ is said to be **weakly measurable** if for each $x^* \in X^*$, the real function $x^* \circ f : S \to \mathbb{R}$ is measurable.

Proposition 1.1.10 *(1)The sets $E(S, X)$ and $F_{sm}(S, X)$, with addition and scalar multiplication pointwise defined, are vector spaces.*

(2) Let u be a function from X into a Banach space Y and

$$f = \sum_i \chi_{A_i} \otimes x_i,$$

an elementary measurable function from S into X, then we have:

$$u \circ f = \sum_i \chi_{A_i} \otimes u(x_i).$$

so that $u \circ f$ is elementary with values in Y. This gives the proof of the following:

Proposition 1.1.11. *If u is a continuous function from X into a Banach space Y and if $f : S \to X$ is strongly measurable, then $u \circ f$ is strongly measurable. In particular the function $s \to \|f(s)\|$ is measurable in the usual sense.*

Proof: Let $\{s_n\}$ be a sequence of elementary measurable functions converging $\mu-$almost everywhere to f. Then $u(s_n)$ is elementary by the preceding Proposition and $u(s_n)$ converges $\mu-$almost everywhere to $u \circ f$ by the continuity of u. To see that $s \to \|f(s)\|$ is measurable take $u(\cdot) = \|\cdot\|$. $\square$

As for the relation between the two types of measurability, weak-strong, this is given by the following theorem of **Pettis**:

Theorem 1.1.12. *A function $f : S \to X$ is strongly measurable if and only if the following conditions are satisfied:*

(a) f is weakly measurable

(b) There is a set $S_0 \in \mathcal{A}$ such that $\mu(S \setminus S_0) = 0$ and the image $f(S_0)$ of S_0 by f is separable. We say that f is $\mu-$almost separably valued. In particular, if X is a separable Banach space, the weak and strong measurability are equivalent.

Proof: ($\Rightarrow$) Let f be strongly measurable.

To see (a), take $u = x^*$, for $x^* \in X^*$, in **Proposition 1.1.11**.

To prove (b), let f_n be elementary converging $\mu-$almost everywhere to f. So let $S_0 \in \mathcal{A}$ be such that $\mu(S \setminus S_0) = 0$ and $f_n(s) \longrightarrow f(s)$, $\forall s \in S_0$. We prove that $f(S_0)$ is separable. Put $A_n = f_n(S_0)$, then A_n is countable since f_n is elementary, and so $A = \bigcup_n A_n$ is also countable. We deduce that the closure A^- of A is separable. Since $f_n(s) \longrightarrow f(s)$, $\forall s \in S_0$, we have $f(S_0) \subseteq A^-$, and then $f(S_0)$ is separable.

($\Leftarrow$) Assume (a) and (b). We can suppose X separable, otherwise we replace it by the closed subspace generated by $f(S_0)$ which is separable by (b). Let $\{z_n,\ n = 1, 2, ...\}$ be a countable dense set in X. Then the family of balls $\{B\left(z_j, \frac{1}{n}\right),\ \ j = 1, 2, ...\}$. covers X for each n. Moreover we have:

$$f^{-1}\left(B\left(z_j, \frac{1}{n}\right)\right) = \left\{s \in S:\quad \|f(s) - z_j\| < \frac{1}{n}\right\} \in \mathcal{A},$$

by the measurability of the function $s \to \|f(s) - z_j\|$. Now form the disjoint sets

$$C_{j,n} = f^{-1}\left(B\left(z_i, \frac{1}{n}\right)\right) \setminus \bigcup_{i<j} f^{-1}\left(B\left(z_i, \frac{1}{n}\right)\right)$$

making a partition of S and define for each n the function $f_n : S \to X$ by $f_n(s) = z_{j_n}$, where j_n is the unique j such that $s \in C_{j,n}$. Then (f_n) is a sequence of elementary measurable functions satisfying $\|f_n(s) - f(s)\| < \frac{1}{n}$. So f_n converges (uniformly on S) to f, consequently f is strongly measurable. $\square$

By the way we proved the following lemma:

Lemma: *Let X be a separable Banach. Then any function $f : S \to X$ is the uniform limit of a sequence of elementary functions.*

1.2 Integration of scalar-valued functions

Now let X be a Banach space, and let $\mu : \mathcal{A} = \mathcal{A}(S) \to X$ be a vector measure. If $f : S \to \mathbb{R}$ is a real measurable function, we will define the integral of f with respect to vector measure μ and give some of its properties needed in integral representation. First we consider simple functions.

Definition 1.2.1. Let $f(\bullet) = \sum_{i=1}^{n} \alpha_i \chi_{A_i}(\bullet)$ be a simple measurable function. The **integral** *of* f with respect to μ over the set $E \in \mathcal{A}$ is defined by:

$$\int_E f\, d\mu = \sum_{i=1}^{n} \alpha_i\, \mu\left(E \cap A_i\right).$$

Just as in the customary real case, this integral does not depend on the representation of f.

It is clear that the integral so defined is linear as a function of f, and

$\sigma-$additive as a set function of E. Moreover if $M = \sup_{s \in E} |f(s)|$, then:

$$\left\| \int_E f \, d\mu \right\| = \left\| M \sum_{i=1}^{n} \left(\frac{\alpha_i}{M} \right) \mu(E \cap A_i) \right\|$$

$$\leq M \left\| \sum_{i=1}^{n} \frac{\alpha_i}{M} \mu(E \cap A_i) \right\| \leq M \, \|\mu\|(E),$$

so we deduce that:

$$\left\| \int_E f \, d\mu \right\| \leq \left(\sup_{s \in E} |f(s)| \right) \|\mu\|(E). \qquad (\#1.2)$$

Definition 1.2.2. A measurable function $f : S \to \mathbb{R}$ is said to be $\mu-$**integrable**, if there is a sequence f_n of simple functions such that:

(a) f_n converges to f $\mu-$almost everywhere

(b) The sequence $\left\{ \int_E f_n \, d\mu \right\}$ converges in the norm of X for each $E \in \mathcal{A}$.

The limit of the sequence $\left\{ \int_E f_n \, d\mu \right\}$ in (b) is called the **integral of** f **with respect to** μ **over** E and is denoted by $\int_E f \, d\mu$.

The integral so defined does not depend on the sequence f_n chosen. This fact is not trivial at all (as it involves applications of **Vitali-Hahn-Saks Theorem** and **Egoroff Theorem**; see [**DS58**, IV.10.8, p. 323-324]). On the other hand, it is straightforward that the integral $\int_E f \, d\mu$ is linear in f.

We record some properties of this integral in the following theorem:

Theorem 1.2.3. (a). *If f is $\mu-$essentially bounded on the set E, then f is $\mu-$integrable over E and :*

$$\left\| \int_E f \, d\mu \right\| \leq \left(\mu - ess \sup_{s \in E} |f(s)| \right) \cdot \|\mu\|(E).$$

(b). *Let T be a linear bounded operator from X into the Banach space Y. Then $T\mu$ is a $Y-$valued vector measure on $\mathcal{A}$, and for any $\mu-$integrable* *f and any $E \in \mathcal{A}$ we have*

$$T\left(\int_E f \, d\mu\right) = \int_E f \, dT\mu.$$

Proof: It is easy to see that the conclusions are valid for f simple (see the inequality $(\#1.2)$ in **Definition 1.2.1**). The fact that $T\mu$ is a vector measure comes from the boundedness of T.

Now let f be measurable with μ–essential bound B on E. Let $\varepsilon > 0$ and let $F_1, F_2, ..., F_n$ be a covering of $f(E)$ by disjoint Borel sets of scalars. Define $E_j = f^{-1}(F_j)$. Let $\alpha_j \in F_j$ and define $f_\varepsilon(s) = \alpha_j$ for $s \in E_j$. Then f_ε is simple and we can arrange matters so that

$$\mu - ess \sup_{s \in E} |f_\varepsilon(s) - f(s)| < \varepsilon.$$

Let $\varepsilon_n \to 0$, we have $\lim_{m,n} \mu - ess \sup_{s \in E} |f_{\varepsilon_n}(s) - f_{\varepsilon_m}(s)| = 0$. By the inequality $(\#1.2)$ in **Definition 1.2.1**, we deduce that

$$\lim_{m,n} \left| \int_E f_{\varepsilon_n} d\mu - \int_E f_{\varepsilon_m} d\mu \right| = 0.$$

This yields the convergence of $\int_E f_{\varepsilon_n} d\mu$ for each $E \in \mathcal{A}$ and we have

$$\int_E f \, d\mu = \lim_n \int_E f_{\varepsilon_n} d\mu.$$

On the other hand, since

$$|f_{\varepsilon_n}| \leq |f_{\varepsilon_n}(s) - f(s)| + |f(s)|,$$

we deduce that

$$\mu - ess \sup_{s \in E} |f_{\varepsilon_n}(s)| \leq B + \varepsilon_n,$$

so that the validity of (a) follows from its validity for simple functions (see the inequality $(\#1.2)$ in definition **1.2.1**). Part (b) comes from its trivial validity for simple functions and the boundedness of T. $\square$

1.3 Bochner Integral of Vector-valued Functions

Let $(S, \mathcal{A}, \mu)$ be a measure space with μ a finite positive measure. We will assume that $(S, \mathcal{A}, \mu)$ is complete. As considered in **Sections 1.1, 1.2,** X

will be a Banach space with topological dual X^*. In this section we define the Bochner integral with respect to the scalar measure μ, for functions $f : S \to X$. (for all details on Bochner integral, see [**HP57**]).

Definition 1.3.1. We say that the elementary measurable function $f(\bullet) = \sum_i x_i \cdot \chi_{A_i}(\bullet)$ is $\mu-$**integrable** if

$$\sum_i \|x_i\| \cdot \mu(A_i) < \infty.$$

In this case we define the integral of f with respect to μ by

$$\int_S f \, d\mu = \sum_i x_i \cdot \mu(A_i).$$

Likewise the integral of f over the set $E \in \mathcal{A}$ is

$$\int_E f \, d\mu = \sum_i x_i \cdot \mu(A_i \cap E).$$

Proposition 1.3.2. *The integral of an elementary measurable function has the following properties:*

(a) If f, g are elementary $\mu-$integrable and if α, β are scalars then $\alpha.f + \beta.g$ is elementary $\mu-$integrable and

$$\int_E (\alpha.f + \beta.g) \, d\mu = \alpha \int_E f \, d\mu + \beta \int_E g \, d\mu.$$

(b) If f is elementary $\mu-$integrable, then

$$\left\| \int_E f \, d\mu \right\| \leq \int_E \|f\| \, d\mu,$$

where $\|f(\bullet)\| = \sum_i \|x_i\| \cdot \chi_{A_i}(\bullet)$.

(c) If Y is a Banach space and if $T : X \to Y$ is a linear bounded operator, then for each elementary $\mu-$integrable function f, the function Tf is elementary $\mu-$integrable and we have

$$T \left(\int_E f \, d\mu \right) = \int_E Tf \, d\mu.$$

Proof: To see (a), write $f = \sum_n s_n \cdot \chi_{A_n}$, and $g = \sum_k t_k \cdot \chi_{B_k}$. Then

$$f + g = \sum_{n,k} (s_n + t_k) \chi_{A_n \cap B_k} \quad \text{and} \quad \alpha f = \sum_n \alpha s_n \cdot \chi_{A_n}.$$

This implies that

$$\|f + g\| = \sum_{n,k} \|s_n + t_k\| \, \chi_{A_n \cap B_k} \leq \sum_{n,k} (\|s_n\| + \|t_k\|) \, \chi_{A_n \cap B_k}$$

$$= \sum_{n} \|s_n\| \cdot \chi_{A_n} + \sum_{k} \|t_k\| \cdot \chi_{B_k}.$$

So we deduce that

$$\sum_{n,k} \|s_n + t_k\| \, \mu\left(A_n \cap B_k\right) \leq \sum_{n} \|s_n\| \, \mu(A_n) + \sum_{k} \|t_k\| \, \mu\left(B_k\right) < \infty.$$

Consequently $f + g$ is elementary μ–integrable and

$$\int_E (f + g) \, d\mu = \int_E f \, d\mu + \int_E g.d\mu.$$

Likewise αf is elementary μ–integrable and

$$\int_E \alpha f \, d\mu = \alpha \int_E f \, d\mu.$$

Part (b) is trivial. To prove (c), we have $Tf = \sum_{n} Ts_n \cdot \chi_{A_n}$ and

$$\int Tf \, d\mu = \sum_{n} Ts_n \cdot \mu(A_n) = T\left(\sum_{n} s_n \cdot \mu(A_n)\right) = T\left(\int_E f \, d\mu\right),$$

by the boundedness of T. $\square$

Remark: If we take in (c) $Y = \mathbb{R}$ and $T = x^*$ for some $x^* \in X^*$ we get

$$x^*\left(\int_E f \, d\mu\right) = \int_E x^* f \, d\mu.$$

Definition 1.3.3. A function $f : S \rightarrow X$ is said to be μ–**integrable** if there is a sequence f_n of elementary μ–integrable functions such that:

(1). f_n converges μ–almost everywhere to f.

(2). $\lim_n \int_E \|f_n - f\| \, d\mu = 0$

In this case the Bochner integral of f is defined by

$$\int_E f \, d\mu = \lim_n \int_E f_n \, d\mu.$$

The following observations legitimate this definition:

First f is strongly measurable by **(1)**. Next the function $\|f_n - f\|$ is positive measurable by **Proposition 1.1.11**, so the integrals $\int_E \|f_n - f\|\, d\mu$ make sense. Finally the sequence $\int_E f_n\, d\mu$ is Cauchy in the Banach space X. For we have

$$\left\|\int_E f_n\, d\mu - \int_E f_m\, d\mu\right\| = \left\|\int_E (f_n - f_m)\, d\mu\right\| \leq \int_E \|(f_n - f_m)\|\, d\mu$$

by **Proposition 1.3.2 (b)**, since $f_n - f_m$ is elementary; so we get

$$\left\|\int_E f_n\, d\mu - \int_E f_m\, d\mu\right\| \leq \int_E \|f_n - f\|\, d\mu + \int_E \|f_m - f\|\, d\mu \to 0, n, m \to \infty,$$

by **(2)**.

Now, if f_n, g_n are two sequences of elementary μ–integrable functions satisfying $(1)-(2)$ we have

$$\left\|\int_E f_n\, d\mu - \int_E g_n\, d\mu\right\| \leq \int_E \|(f_n - g_n)\|\, d\mu$$

$$\leq \int_E \|f_n - f\|\, d\mu + \int_E \|g_n - f\|\, d\mu \to 0, n \to \infty.$$

Consequently the Bochner integral of f is well defined. The following theorem gives one of the outstanding facts about the Bochner integral.

Proposition 1.3.4: For every μ–integrable function $f : S \to X$ and every $x^* \in X^*$ we have

$$x^*\left(\int_E f\, d\mu\right) = \int_E x^* f\, d\mu.$$

Proof: Let $\{f_n\}$ be a sequence of elementary functions defining $\int_E f\, d\mu$.

Since $f_n \to f$, $\mu.a.e$ and since x^* is continuous, we have $x^* f_n \to x^* f$, $\mu.a.e.$

On the other hand $x^*\left(\int_E f_n\, d\mu\right) = \int_E x^* f_n\, d\mu$ by **Proposition 1.3.2 (c)**.

We deduce

$$\left| x^* \left(\int_E f_n \, d\mu \right) - \int_E x^* f \, d\mu \right| = \left| \int_E x^* f_n \, d\mu - \int_E x^* f \, d\mu \right|$$

$$\leq \int_E |x^* f_n - x^* f| \, d\mu \leq \|x^*\| \int_E \|f_n - f\| \, d\mu \to 0.$$

But $\displaystyle\int_E f_n \, d\mu \to \int_E f \, d\mu$, so

$$x^* \left(\int_E f_n \, d\mu \right) \to x^* \left(\int_E f \, d\mu \right).$$

Consequently $\displaystyle x^* \left(\int_E f \, d\mu \right) = \int_E x^* f \, d\mu.$ $\square$

Theorem 1.3.5.(Bochner) *A function* $f : S \to X$ *is* $\mu-$*integrable if and only if* f *is strongly measurable and* $\int_S \|f\| \, d\mu < \infty.$

In other words, for a strongly measurable function f, *the* $\mu-$*integrability is equivalent to the integrability of* $\|f\|.$

Proof: Suppose f $\mu-$integrable. There exists a sequence (f_n) of $\mu-$integrable elementary functions such that $f_n \to f$, $\mu.a.e$ and $\displaystyle\int_E \|f_n - f\| \, d\mu \to 0$; in particular f is strongly measurable and

$$\int_S \|f\| \, d\mu \leq \int_S \|f_n - f\| \, d\mu + \int_S \|f_n\| \, d\mu < \infty.$$

Conversely let f be strongly measurable and satisfies

$$\int_S \|f\| \, d\mu < \infty.$$

By **Definition 1.3.3**, we have to show the existence of a sequence g_n of elementary $\mu-$integrable functions such that g_n converges $\mu-$almost everywhere to f and $\lim_n \displaystyle\int_E \|g_n - f\| \, d\mu = 0$. Fix $\alpha > 0$ and let us consider the sets

$$K_n = \{s \in S, \quad \|f_n(s)\| \leq \|f(s)\| \cdot (1 + \alpha)\}.$$

Now define the function g_n by $g_n = f_n \chi_{K_n}$. Since $\|f_n\|$ and $\|f\|$ are measurable, it results that $K_n \in \mathcal{A}$ and so g_n is elementary measurable. On the otherhand we have

$$\|g_n\| = \|f_n\| \chi_{K_n} \leq \|f\| \cdot (1 + \alpha) \quad \text{on all of } S.$$

From the condition $\int_S \|f\|\,d\mu < \infty$ we deduce that g_n is μ−integrable. Since $f_n \to f$, $\mu.a.e$, we have: for almost every $s \in S$, there exists $N(s) \geq 1$ such that

$$\|f_n(s)\| - \|f(s)\| \leq \alpha \|f(s)\| \; \forall n \geq N(s);$$

that is $s \in K_n$. This means that $g_n(s) = f_n(s)\chi_{K_n}(s) = f_n(s)$, $\forall n \geq N(s)$. This proves that g_n converges μ−almost everywhere to f and then $\|g_n - f\| \to 0$ $\mu.a.e$. Since we have

$$\|g_n - f\| \leq \|g_n\| + \|f\| \leq \|f\| \cdot (1 + \alpha) + \|f\| = \|f\| \cdot (2 + \alpha),$$

we deduce from dominated convergence theorem in the classical space $L_1(\mu)$ that

$$\lim_n \int_E \|g_n - f\|\,d\mu = 0. \;\square$$

We denote by $L_1(\mu, X)$ the set of all μ−integrable functions.

As usual we identify two functions that are equal μ−almost everywhere, in symbols $f = g$ $\mu - a.e.$ it easy to check that:

(a) $L_1(\mu, X)$ is a vector space and the map $f \to \int_E f\,d\mu$ is linear;

(b) $f = g$ $\mu - a.e \iff \int_E f\,d\mu = \int_E g\,d\mu$, for all $E \in \mathcal{A}$.

Now we extend the definition of the Bochner integral to a real measure μ with bounded variation denoted by $|\mu|$. First write $\mu = \mu^+ - \mu^-$ and $|\mu| = \mu^+ + \mu^-$, where μ^+, μ^- are the positive and negative parts of μ.

Definition 1.3.10. A function $f : S \to X$ is said to be μ−**integrable** if it is μ^+−integrable and μ^-−integrable. In this case the **Bochner integral** of f with respect to the real measure μ is defined by

$$\int_E f\,d\mu = \int_E f\,d\mu^+ - \int_E f\,d\mu^-.$$

For any real μ, it is customary to denote by $L_1(\mu, X)$ the vector space of all equivalence classes, with respect to the equality $|\mu|-a.e.$, of μ−integrable functions. If $f \in L_1(\mu, X)$ we put

$$\|f\|_1 = \int_S \|f\|\,d\mu.$$

Then we have:

Theorem 1.3.11. $\|\bullet\|_1$ *is a norm which makes $L_1(\mu, X)$ a Banach space and we have*

$$\left\| \int_E f\,d\mu \right\| \leq \int_E \|f\|\,d\mu, \text{ for all } E \in F \text{ and } f \in L_1(\mu, X).$$

Moreover the operator $I_\mu : L_1(\mu, X) \to X$ *given by* $I_\mu(f) = \int_S f\, d\mu$ *is linear continuous with norm* $\|I_\mu\| = |\mu|$, $|\mu|$ being the total variation of μ.

Proof: Mimic the classical proof for the real space $L_1(\mu)$. Let $\{f_n\}$ be Cauchy in $L_1(\mu, X)$, so $\int_S \|f_n - f_m\|\, d\mu \to 0$, $n, m \to \infty$. It is not difficult to find a sequence of integers $\{n_i\}$ such that $1 \leq n_i < n_{i+1}$, and

$$\int_S \|f_{n_{i+1}} - f_{n_i}\|\, d\mu < 2^{-i} \text{ for all } i \geq 1.$$

We deduce that

$$\int_S \|f_{n_1}\|\, d\mu + \sum_i \int_S \|f_{n_{i+1}} - f_{n_i}\|\, d\mu < \infty.$$

Therefore the function $\|f_{n_1}\| + \sum_i \|f_{n_{i+1}} - f_{n_i}\|$ is finite $\mu - a.e.$ Let

$$S_0 = \left\{ s : \|f_{n_1}(s)\| + \sum_i \|(f_{n_{i+1}} - f_{n_i})(s)\| < \infty \right\}, \mu(S \backslash S_0) = 0$$

and we can define:

$$f(s) = f_{n_1}(s) + \sum_i (f_{n_{i+1}} - f_{n_i})(s), \text{ if } s \in S_0,$$

$$\text{and } f(s) = 0 \text{ if } s \in S \backslash S_0.$$

Then it is clear that $f_{n_i}(s) \to f(s)$ for all $s \in S_0$, this proves that f is strongly measurable. On the other hand

$$\int_S \|f - f_{n_i}\|\, d\mu \leq \sum_{i+1}^{\infty} \int_S \|f_{n_{i+1}} - f_{n_i}\| \to 0, i \to \infty,$$

so $\int_S \|f\|\, d\mu. < \infty$, and f is μ–integrable by **Theorem 1.3.9**. Now we have

$$\int_S \|f_n - f\|\, d\mu \leq \int_S \|f - f_{n_i}\|\, d\mu + \int_S \|f_n - f_{n_i}\|\, d\mu \to 0, n, i \to \infty.$$

To see the relation $\left\| \int_E f\, d\mu \right\| \leq \int_E \|f\|\, d\mu$ and the continuity of I_μ use the

Definition 1.3.3 of $\int_E f\, d\mu$. $\square$

We close this section with the dominated convergence theorem in $L_1(\mu, X)$:

Theorem 1.3.12. Let (f_n) be a sequence in $L_1(\mu, X)$ such that $f_n \to f$ $\mu - a.e.$ Suppose there exists $g \in L_1(\mu)$ such that for each n we have $\|f_n\| \leq g$, $\mu - a.e.$ Then $f \in L_1(\mu, X)$ and $\int_S f\, d\mu = \lim_n \int_S f_n\, d\mu.$

Proof: For each $x^* \in X^*$ the scalar functions $x^* f_n$ converge to $x^* f$ $\mu - a.e.$, so f is weakly measurable. On the other hand, since f_n is strongly measurable $f_n(S)$ is separable. We deduce that $\bigcup_n f_n(S)$ is separable. Since $f(S) \subset \overline{\bigcup_n f_n(S)}$, it follows that $f(S)$ is separable. Now we apply **Pettis Theorem 1.3.4**.toget that f is strongly measurable. Since $\|f_n\| \leq g$, $\mu - a.e.$ we deduce $\|f\| \leq g$, $\mu - a.e.$and then $\int_S \|f\| \, d\mu < \infty$, consequently f is Bochner integrable by **Theorem 1.3.9**, so $f \in L_1(\mu, X)$. On the other hand,

$$\|f_n - f\| \leq 2g \quad \text{and} \quad 2g - \|f_n - f\| \geq 0.$$

By **Fatou lemma**

$$
\begin{aligned}
0 \;\; &\leq \;\; \int_S \liminf_n 2g - \|f_n - f\| \, d\mu = \int_S 2g \, d\mu \\
&\leq \;\; \liminf_n \int_S (2g - \|f_n - f\|) \, d\mu \\
&= \;\; \int_S 2g \, d\mu - \limsup_n \int_S \|f_n - f\| \, d\mu.
\end{aligned}
$$

It follows that $\limsup_n \int_S \|f_n - f\| \, d\mu \leq 0$, so $\lim_n \int_S \|f_n - f\| \, d\mu = 0.\square$

1.4   Operator Valued Measures

Let X, Y be Banach spaces, S a compact space and let $C(S, X)$ be the Banach space of all continuous functions $f : S \to X$, with the sup. norm:

$$\|f\| = \sup_{s \in S} \|f(s)\|, f \in C(S, X).$$

Put $C(S, X) = C(S)$ if $X = \mathbb{R}$.

The symbol E^* means, as usual, the topological dual of the Banach space E and E^{**} its bidual. We will denote by $L(X, E)$ the space of all linear bounded operators from X into the Banach space E.

All the set functions considered here are, except otherwise stated, assumed to be defined on the Borel $\sigma-$algebra $\mathcal{B}(S)$ of S.

Definition 1.4.1. We will deal with additive operator valued set functions $G : \mathcal{B}(S) \to L(X, E)$, which we will call **operator valued measures**.

Definition 1.4.2 The **semivariation of an operator valued measure** $G : \mathcal{B}(S) \to L(X, E)$ is defined by:

$$\widetilde{G}(B) = \sup \left\| \sum_i G(A_i).x_i \right\|, B \in \mathcal{B}(S),$$

the supremum being taken over all finite partitions $\{A_i\}$ of B in $\mathcal{B}(S)$ and all finite systems of vectors $\{x_i\}$ in X, with $\|x_i\| \leq 1$, $\forall i$.

Definition 1.4.3. If $\lambda : \mathcal{B}(S) \to E$ is a vector measure, the **variation of** λ is defined by the positive set function $|\lambda|\,(\bullet)$ given by:

$$|\lambda|\,(B) = \sup \sum_i \|\lambda(A_i)\|\,,\ B \in \mathcal{B}(S).$$

the supremum being taken over all finite partitions $\{A_i\}$ of B in $\mathcal{B}(S)$. Let us observe that this is exactly **Definition 1.1.4.**

We say that λ is of bounded variation if $|\lambda|\,(B) < \infty$, for all $B \in \mathcal{B}(S)$ and in this case, $|\lambda|$ itself is a finite positive measure. λ is said to be regular if $|\lambda|$ is regular in the customary sense (see [**DU77**] for details).

Let us denote by $M_{r\sigma bv}(\mathcal{B}(S), E)$ the vector space of all regular $E-$valued vector measures with bounded variation, put $M_{r\sigma bv}(\mathcal{B}(S), E) = M_{r\sigma bv}(\mathcal{B}(S))$, if $E = \mathbb{R}$. The variation $|\lambda|$ defines a norm on the space $M_{r\sigma bv}(\mathcal{B}(S), E)$, and for this norm, $M_{r\sigma bv}(\mathcal{B}(S), E)$ is a Banach space. (see [**DS58, Chapter III**] for details). Moreover, we have:

Theorem 1.4.4. *In the specific case $E = X^*$*

$$|\lambda|\,(B) = \sup |\sum_i \lambda(A_i).x_i|, B \in \mathcal{B}(S),$$

the supremum being taken over all finite partitions $\{A_i\}$ of B in $\mathcal{B}(S)$ and all finite systems of vectors $\{x_i\}$ in X, with $\|x_i\| \leq 1$, $\forall i$.

In other words, for X^*-valued vector measures, the variation is equal to the semivariation (see **Definition 1.4.2**).

Proof: We have

$$|\sum_i \lambda(A_i).x_i| \leq \sum_i |\lambda(A_i).x_i| \leq \sum_i \|\lambda(A_i)\|\,, if\ \|x_i\| \leq 1,\ \forall i,$$

taking appropriate supremum we get $\widetilde{\lambda}(B) \leq |\lambda|\,(B)$. To see the reverse inequality, let (A_i) be a finite partition of B in $\mathcal{B}(S)$, $1 \leq i \leq n$. If $\varepsilon > 0$ is given then for each $i \leq n$ there is x_i with $\|x_i\| = 1$, and

$$\|\lambda(A_i)\| \leq \lambda(A_i).x_i + \frac{\varepsilon}{n}.$$

We get

$$\sum_i \|\lambda(A_i)\| \leq \sum_i \lambda(A_i).x_i + \varepsilon = |\sum_i \lambda(A_i).x_i| + \varepsilon.$$

By taking appropriate supremum, this yields

$$|\lambda|\,(B) \leq \widetilde{\lambda}(B) + \varepsilon,$$

so $|\lambda|\,(B) \leq \widetilde{\lambda}(B)$, since ε is arbitrary. $\square$

Definition 1.4.5. To each additive set function $G : \mathcal{B}(S) \to L\left(X, Y^{**}\right)$, there is an **associated a family** $\{G_{y^*} : y^* \in Y^*\}$ **of additive** X^*-**valued set functions** given by:

$$G_{y^*}(A).x = G(A).x\,(y^*), \quad A \in \mathcal{B}(S).$$

Note that since $G(A) \in L\left(X, Y^{**}\right)$, we have:

$$\|G_{y^*}(A).x\| \le \|G(A).x\| \cdot \|y^*\| \le \|G(A)\| \cdot \|x\| \cdot \|y^*\|.$$

Moreover we have the important estimation :

Proposition 1.4.6. *For any additive* $G : \mathcal{B}(S) \to L\left(X, Y^{**}\right)$, *we have:*

$$\widetilde{G}(A) = \sup\{|G_{y^*}|\,(A),\ \|y^*\| \le 1\}, A \in \mathcal{B}(S).$$

Proof: Since G_{y^*} is X^*-valued, we have $\widetilde{G_{y^*}} = |G_{y^*}|\,(A)$. Now for every finite partition $\{A_i\}$ of A in $\mathcal{B}(S)$, and every finite systems of vectors $\{x_i\}$ in X, with $\|x_i\| \le 1$, $1 \le i \le n$, we have

$$
\left\| \sum_i G(A_i).x_i \right\| = \sup_{\|y^*\| \le 1} \left| \left(\sum_i G(A_i).x_i \right)(y^*) \right|
$$

$$
= \sup_{\|y^*\| \le 1} \left| \sum_i G(A_i).x_i\,(y^*) \right|
$$

$$
= \sup_{\|y^*\| \le 1} \left| \sum_i G_{y^*}(A_i).x_i \right|
$$

$$
\le \sup_{\|y^*\| \le 1} \widetilde{G_{y^*}}(A) = \sup_{\|y^*\| \le 1} |G_{y^*}|\,(A).
$$

Taking the supremum of $\left\| \sum_i G(A_i).x_i \right\|$ over all finite partitions $\{A_i\}$ of A in $\mathcal{B}(S)$, and all finite systems of vectors $\{x_i\}$ in X, with $\|x_i\| \le 1$, we get $\widetilde{G}(A) \le \sup_{\|y^*\| \le 1} |G_{y^*}|\,(A)$. On the other hand if $y^* \in Y^*, \|y^*\| \le 1$ then

$$
\left| \sum_i G_{y^*}(A_i).x_i \right| = \left| \sum_i G(A_i).x_i \right)(y^*) \right|
$$

$$
\le \left\| \sum_i G(A_i).x_i \right\| \le \widetilde{G}(A),
$$

therefore

$$
\widetilde{G_{y^*}}(A) \le \widetilde{G}(A),
$$

$$
\text{and} \quad \sup_{\|y^*\| \le 1} \widetilde{G_{y^*}}(A) = \sup_{\|y^*\| \le 1} |G_{y^*}|\,(A) \le \widetilde{G}(A). \qquad \square
$$

1.5   Integral of Vector-valued Functions

Now the aim is to define the integral of a function f on S with values in X, against an additive $L(X, E)-$valued set function G on $\mathcal{B}(S)$. The result of this integration will be a vector of the Banach space E. First, we define the measurable functions which will be integrated.

Definition 1.5.1. Let $f : S \to X$ be a function on S with values in the Banach space X.

(1) f is **simple measurable** if it has the form

$$f(\bullet) = \sum_{i=1}^{n} x_i \cdot \chi_{A_i}(\bullet),$$

where $x_i \in X$, $A_i \in \mathcal{B}(S)$, $i = 1, 2, ..., n$, the A_i being pairwise disjoint.

(2) f is **measurable** if it is the uniform limit of a sequence of simple measurable functions.

We denote by $F_m(S, X)$ the set of all measurable functions $f : S \to X$. It is clear that $F_m(S, X)$ is a vector space. Moreover, in the Banach space of all bounded functions from S into X, with the uniform norm, $F_m(S, X)$ is the closure of the subspace of all simple measurable functions.

Definition 1.5.2. Let G be an additive $L(X, E)-$valued set function on $\mathcal{B}(S)$. If

$$f(\bullet) = \sum_{i=1}^{n} x_i \cdot \chi_{A_i}(\bullet)$$

is a simple measurable function, the **integral** of f on the set $B \in \mathcal{B}(S)$ with respect to G is defined by the formula:

$$\int_B f \, dG = \sum_{i=1}^{n} G(A_i \cap B) . x_i$$

From the additivity of G, it is easy to see that the integral does not depend on the form of f, and is linear. Moreover if $M = \sup_{s \in B} \|f(s)\|$, then:

$$\left\| \int_B f \, dG \right\| = M \left\| \sum_{i=1}^{n} G(A_i \cap B) . \frac{x_i}{M} \right\| \leq M . \widetilde{G}(B).$$

So we get for each simple f and each $B \in \mathcal{B}(S)$:

$$\left\| \int_B f \, dG \right\| \leq \left(\sup_{s \in B} \|f(s)\| \right) . \widetilde{G}(B), \qquad (\#1.3)$$

where $\widetilde{G}(B)$ is the semivariation defined in **1.4.2**.

Definition 1.5.3. Let $f : S \to X$ be a measurable function, i.e $f \in F_m(S, X)$. We define the **integral of f over B with respect to** G by the limit,

$$\int_B f \, dG = \lim_n \int_B f_n \, dG,$$

where f_n is any sequence of simple measurable functions converging uniformly to f. The limit is in the norm of X.

Remark. To check that the integral is well defined, let $\{f_n\}, \{g_n\}$ be two sequences of simple measurable functions converging uniformly to f. Then by the inequality $(\#\mathbf{1.3})$, we have:

$$\left\| \int_B f_n \, dG - \int_B g_n \, dG \right\| = \left\| \int_B (f_n - g_n) \, dG \right\|$$

$$\leq \left(\sup_n \| f_n(s) - g_n(s) \| \right) \widetilde{G}(B) \longrightarrow 0, \quad n \longrightarrow \infty.$$

Proposition 1.5.4. *For each measurable function f and each $B \in \mathcal{B}(S)$, we have:*

$$\left\| \int_B f \, dG \right\| \leq \left(\sup_{s \in B} \| f(s) \| \right) . \widetilde{G}(B).$$

Proof: Let f_n be a sequence of simple measurable functions converging uniformly to f. If $\varepsilon > 0$, there is an integer $N \geq 1$ such that for all $n \geq N$, we have

$$\left\| \int_B f \, dG - \int_B f_n \, dG \right\| < \varepsilon.$$

So we deduce that for all $n \geq N$:

$$\left\| \int_B f \, dG \right\| \leq \left\| \int_B f \, dG - \int_B f_n \, dG \right\| + \left\| \int_B f_n \, dG \right\|$$

$$< \varepsilon + \left(\sup_{s \in B} \| f_n(s) \| \right) . \widetilde{G}(B).$$

Letting $n \longrightarrow \infty$ and then $\varepsilon \searrow 0$, we get the inequality. $\square$

In some representation theorems, we need to integrate continuous functions $f : S \to X$. This will be possible according to the following proposition. Recall that $C(S, X)$ is the Banach space of all continuous functions $f : S \to X$, with the sup. norm.

Proposition 1.5.5. *The Banach space* $C(S, X)$ *is a closed subspace of* $F_m(S, X)$.

Proof: Let $\varphi = f \otimes x \in C(S) \otimes X$. Since f is bounded, there is a sequence s_n of scalar simple measurable functions converging uniformly to f on S. This yields the sequence $f_n = s_n \otimes x$ of $X-$valued simple measurable functions converging uniformly to φ on S. So such φ are in $F_m(S, X)$. By **Theorem A.5.7** (or see also [**Din67, Proposition 1, § 19**], $C(S) \otimes X$ is dense in $C(S, X)$. Since $F_m(S, X)$ is closed, we get $C(S, X) \subseteq F_m(S, X)$. $\square$

1.6   Notes and comments

The general setting of an integration process involves two main mathematical objects, a function and a measure. The function may be scalar or vector valued, likewise for the measure which may even be operator valued. This chapter describes three integration processes, according to their future use in integral representation. In the first one scalar functions are integrated with respect to vector measures with values in a Banach space. This is the basic process used by Bartle-Dunford-Scwartz in integral representation [**BDS55, DS58**]. The second concerns the Bochner integral of a Banach valued function with respect to a real measure. This process is fully described in [**HP57**]. It is used by L. Meziani [**Mez02**] to characterize some classes of bounded operators by they Bochner integral form. The third process considered that of a vector function with respect to an operator valued measure given togather with one of the most general integral representation theorem in [**Din67**]. For an exhaustive presentation of vector measures and integration the reader is refered to [**DU77**].

2

Characterization of Duals of Function Spaces

2.1 The Dual of $(C_0(S), u)$

This section contains characterization of duals of function spaces $(C_0(S), u)$ and $(C_0(S), u)$. This has been treated in several books, for example, by Aliprantis and Burkinshaw [**AB98**], Folland [**Fol99**], Royden [**Roy88**], Rudin [**Rud87**].

Throughout this section, S will denote a locally compact Hausdorff space. We recall that:

(1) $C_c(S)$ is the space of continuous functions on S with compact support.

(2) $C_0(S)$ is the space of continuous functions on S which vanish at infinity.

(3) If U is open in S and $f \in C_c(S)$, we shall write $f \prec U$ to mean that $0 \leq f \leq 1$ and $supp(f) \subseteq U$. (This is slightly stronger than the condition $0 \leq f \leq \chi_U$, which implies only that $supp(f) \subseteq \overline{U}$.)

(4) [**Rud87, p. 37**] Suppose U is open in a locally compact Hausdorff space S, $K \subseteq U$, and K is compact. Then there is an open set V with compact closure such that

$$K \subseteq V \subseteq \overline{V} \subseteq U.$$

(5) **Urysohn's Lemma** (for locally compact spaces) [**Rud87, p. 39**] Suppose S is a locally compact Hausdorff space, V is open in S, $K \subseteq V$, and K is compact. Then there exists an $f \in C_c(S)$, such that

$$K \prec f \prec V$$

In terms of characteristic functions, the conclusion asserts the existence of a continuous function f which satisfies the inequalities $\chi_K \leq f \leq \chi_V$. Note that it is easy to find semicontinuous functions which do this; examples are χ_S and χ_V.

Lemma 2.1.1. (Partition of Unity) [**Rud87, p. 40**] *Let S be a locally compact Hausdorff space, $U_1, \dots, U_n$ open subsets of S and K a compact subset of S such that*

$$K \subseteq U_1 \bigcup \dots \bigcup U_n.$$

Then there exist $\varphi_1, \dots, \varphi_n \in C_c(S)$ such that $\varphi_i \prec U_i$ $(i = 1, 2, \dots, n)$, and

$$\sum_{i=1}^{n} \varphi_i = 1 \text{ on } K \text{ and } \sum_{i=1}^{n} \varphi_i \leq 1 \text{ on } S. \qquad (1)$$

Because of **(1)**, the collection $\{\varphi_1, ..., \varphi_n\}$ is called a partition of unity on K, subordinate to the cover $\{U_1, ..., U_n\}$.

Proof. Let $x \in K$. Since $K \subseteq \bigcup_{i=1}^{n} U_i$, $x \in U_i$ for some $1 \le i \le n$. Since S be a locally compact Hausdorff space, there exists a neighbourhood V_x of x in S such that $\overline{V_x}$ is compact and $\overline{V_x} \subseteq U_i$. Clearly, $\{V_x : x \in K\}$ is an open cover of compact K, and so it has a finite subcover $\{V_{x_j} : 1 \le j \le m\}$ (say) for K. For each $1 \le i \le n$, let

$$K_i = \bigcup\{\overline{V_{x_j}} : \overline{V_{x_j}} \subseteq U_i, 1 \le j \le m\}.$$

Clearly, $K \subseteq \bigcup_{i=1}^{n} K_i$. Further, each K_i is compact and $K_i \subseteq U_i$. Since S is locally compact Hausdorff, it is normal. Since K_i and $S \setminus U_i$ are closed disjoint sets in S, by the Urysohn's lemma, there exist functions $\psi_1, ..., \psi_n \in CB(S)$ such that

$$0 \le \psi_i \le 1, \quad \psi_i = 1 \text{ on } K_i \text{ and } \psi_i = 0 \text{ on } S \setminus U_i, \quad (1 \le i \le 1).$$

Define

$$
\begin{aligned}
\varphi_1 &= \psi_1 \\
\varphi_2 &= (1 - \psi_1)\psi_2 \\
&\quad \cdots\cdots \\
\varphi_n &= (1 - \psi_1)(1 - \psi_2)....(1 - \psi_{n-1})\psi_n \quad (n \ge 2).
\end{aligned}
$$

Clearly, $0 \le \varphi_i \le 1$ and $\varphi_i = 0$ on $S \setminus U_i$ and so supp $\varphi_i \subseteq U_i$. By induction, we can write

$$\sum_{i=1}^{n} \varphi_i = \psi_1 + (1 - \psi_1)\psi_2 + ... + (1 - \psi_1)(1 - \psi_2)....(1 - \psi_{n-1})\psi_n$$

$$= 1 - (1 - \psi_1)(1 - \psi_2)....(1 - \psi_n) = 1 - \prod_{i=1}^{n}(1 - \psi_i)$$

If $x \in S$. Then clearly

$$\sum_{i=1}^{n} \varphi_i(x) = 1 - \prod_{i=1}^{n}[1 - \psi_i(x)] \le 1 - 0 = 1.$$

Let $x \in K$. Since $K \subseteq \bigcup_{i=1}^{n} K_i$, $x \in K_j$ for some $1 \le j \le n$, and so $\psi_j(x) = 1$. Hence

$$\sum_{i=1}^{n} \varphi_i(x) = 1 - (1 - \psi_j(x))\prod_{i \ne j}[1 - \psi_i(x)] = 1 - 0 = 1. \quad \square$$

Representation of Positive Linear functional on $C_c(S)$

Recall that:

(1) $Bo(S)$ denotes the **Borel σ-algebra** on S, that is, the σ-algebra generated by open sets (or equivalently, by closed sets).

(2) Measures on $Bo(S)$ will be called **Borel measures**.

(3) Countable unions (resp. intersections) of closed (resp. open) sets will be called F_σ (resp. G_δ) **sets**.

(4) Let $(S, \mathcal{A}(S), \mu)$ be a measure space. If $\mu(S) < \infty$ (which implies that $\mu(E) < \infty$ for all $E \in \mathcal{A}(S)$ since $\mu(S) = \mu(E) + \mu(E^c)$), μ is called **finite**. If $S = \bigcup_{j=1}^{\infty} E_j$ where $E_j \in \mathcal{A}(S)$ and $\mu(E_j) < \infty$ for all j, μ is called σ**-finite**. More generally, if $E = \bigcup_{j=1}^{\infty} E_j$ where $E_j \in \mathcal{A}(S)$ and $\mu(E_j) < \infty$ for all j, we say that E is of σ-finite measure.

Definition: A linear functional T on $C_c(S)$ will be called **positive** if $T(f) \geq 0$ whenever $f \geq 0$.

Note: (1) Every positive linear functional T is **monotone**: $f \leq g$ implies $T(f) \leq T(g)$. [Since $g - f \geq 0$, $T(f - g) \geq 0$, and so

$$T(g) = T(f) + T(g - f) \geq T(f).]$$

(2) In this definition there is no mention of continuity, but it is worth noting that positivity itself implies a rather strong continuity property.

Proposition 2.1.2. [Fol99] *If T is a positive linear functional on $C_c(S)$, for each compact $K \subseteq S$ there is a constant $C_K > 0$ such that*

$$|T(f)| \leq C_K.\|f\|_\infty \text{ for all } f \in C_c(S) \text{ with } supp(f) \subseteq K.$$

Proof. It suffices to consider real-valued f. Given a compact K, choose $\varphi \in C_c(S, [0, 1])$ such that $\varphi = 1$ on K (Urysohn's lemma). Then if $supp(f) \subseteq K$, we have $|f| \leq \|f\|_\infty \varphi$, that is,

$$\|f\|_\infty \varphi - f > 0 \quad \text{and} \quad \|f\|_\infty \varphi + f > 0$$

.Thus

$$\|f\|_\infty T(\varphi) - T(f) > 0 \quad \text{and} \quad \|f\|_\infty T(\varphi) + T(f) > 0,$$

so that $|T(f)| \leq T(\varphi)\|f\|_\infty$. $\square$

Remark: If μ is a Borel measure on S such that $\mu(K) < \infty$ for every compact $K \subseteq S$, then clearly $C_c(S) \subseteq L^1(\mu)$, so the map $f \to \int f d\mu$ is a positive linear functional on $C_c(S)$. The principal result of this section is that every positive linear functional on $C_c(S)$ arises in this fashion; moreover, one can impose some additional regularity conditions on μ, subject to which μ is unique. These conditions are as follows.

Definition: Let μ be a Borel measure on S and E a Borel subset of S. The measure μ is called:

(i) an **outer regular** on E if

$$\mu(E) = \inf\{\mu(U) : U \supseteq E, U \text{ open}\};$$

(ii) an **inner regular** on E if

$$\mu(E) = sup\{\mu(K) : K \subseteq E, K \text{ compact}\}.$$

Definitions: (1) If μ is outer and inner regular on all Borel sets, μ is called **regular**. It turns out that regularity is a bit too much to ask for when S is not σ-compact, so we adopt the following definition.

(2) A **Radon measure** on S is a Borel measure that is finite on all compact sets, outer regular on all Borel sets, and inner regular on all open sets. We shall later see that Radon measures are also inner regular on all of their σ-finite sets.

We shall denote the space of all Radon measures on S by $M(S)$, and for any $\mu \in M(S)$ we define

$$||\mu|| = |\mu|(S),$$

where $|\mu|$ is the total variation of μ. then $M(S)$ is a vector space and $\mu \to ||\mu||$ is a norm on it.

The Riesz Representation Theorem 2.1.3 [Fol99, p.211] If T is a positive linear functional on $C_c(S)$, there is a unique Radon measure μ on S such that $T(f) = \int f d\mu$ for all $f \in C_c(S)$. Moreover, μ satisfies

$$\mu(U) = \sup\{T(f) : f \in C_c(S), f \prec U\} \text{ for all open } U \subseteq S, \quad \textbf{(A1)}$$

and

$$\mu(K) = \inf\{T(f) : f \in C_c(S), f \geq \chi_K\} \text{ for all compact } K \subseteq S. \quad \textbf{(A2)}$$

Proof. Let us begin by establishing uniqueness. If μ is a Radon measure such that $T(f) = \int f d\mu$ forall $f \in C_c(S)$, and $U \subseteq S$ is open, then clearly $T(f) \leq \mu(U)$ whenever $f \prec U$. On the other hand, if $K \subseteq U$ is compact, by **Urysohn's lemma** there is an $f \in C_c(S)$ such that $f \prec U$ and $f = 1$ on K, whence $\mu(K) \leq \int f d\mu = T(f)\cdot$

Since μ is inner regular on U, it follows that $\textbf{(A1)}$ is satisfied. Thus μ is determined by T on open sets, and hence on all Borel sets because of outer regularity. This argument proves the uniqueness of μ and also suggests how to go about proving existence. Namely, we begin by defining

$$\mu(U) = \sup\{T(f) : f \in C_c(S), f \prec U\}$$

for U open, and we then define $\mu^*(E)$ for an arbitrary $E \subseteq S$ by

$$\mu^*(E) = \inf\{\mu(U) : U \supseteq E, U \text{ open}\}.$$

Clearly $\mu(U) \leq \mu(V)$ if $U \subseteq V$, and hence $\mu^*(U) = \mu(U)$ if U is open.

The outline of the proof is now as follows. We shall establish that

(I). μ^* is an outer measure.

(II). Every open set is μ^*-measurable.

At this point it follows from **Caratheodory's theorem** that every Borel set is μ^*-measurable and that $\mu = \mu^*|Bo(S)$ is a Borel measure. (The notation is consistent because $\mu^*(U) = \mu(U)$ for U open.) The measure μ is outer regular and satisfies **(A1)** by definition. We next show that

(III). μ satisfies **(A2)**.

This clearly implies that μ is finite on compact sets, and inner regularity on open sets also follows easily. [Indeed, if U is open and $\alpha \le \mu(U)$, choose $f \in C_c(S)$ such that $f \prec U$ and $T(f) > \alpha$, and let $K = supp(f)$. If $g \in C_c(S)$ and $g \ge \chi_K$, then $g - f > 0$ and hence $T(g) \ge T(f) \ge \alpha$. But then $\mu(K) > \alpha$ by **(A2)**, so μ is inner regular on U.]

Finally, we prove that

(IV). $T(f) = \int f d\mu$ for all $f \in C_c(S)$.

With this, the proof of the theorem will be complete.

Proof of (I): It suffices to show that if $\{U_j\}$ is a sequence of open sets and $U = \bigcup_{j=1}^{\infty} U_j$, then $\mu(U) \le \sum_{j=1}^{\infty} \mu(U_j)$. Indeed, from this it follows that for any $E \subseteq S$,

$$\mu^*(E) = inf\{\sum_{j=1}^{\infty} \mu(U_j) : U_j \text{ open}, E \subseteq \bigcup_{j=1}^{\infty} U_j\},$$

and the expression on the right defines an outer measure. If $U = \bigcup_{j=1}^{\infty} U_j$, $f \in C_c(S)$, and $f \prec U$, let $K = supp(f)$. Since K is compact, we have $K \subseteq \bigcup_{j=1}^{n} U_j$ for some finite n, so by **Partition of Unity**, there exist $g_1, ..., g_n \in C_c(S)$ with $g_j \prec U_j$ and $\sum_{j=1}^{n} g_j = 1$ on K. But then $f =: \sum_{j=1}^{n} f g_j$ and $f g_j \prec U_j$, so

$$T(f) = \sum_{j=1}^{n} T(f g_j) \le \sum_{j=1}^{n} \mu(U_j) \le \sum_{j=1}^{\infty} \mu(U_j).$$

Since this is true for any $f \prec U$, we conclude that $\mu(U) \le \sum_{j=1}^{\infty} \mu(U_j)$ as desired.

Proof of (II): We must show that if U is open and E is any subset of S such that $\mu^*(E) < \infty$, then

$$\mu^*(E) > \mu^*(E \cap U) + \mu^*(E \backslash U).$$

First suppose that E is open. Then $E \cap U$ is open, so given $E > 0$ we can find $f \in C_c(S)$ such that $f \prec E \cap U$ and $T(f) > \mu(E \cap U) - E$. Also,

$E\backslash(supp(f))$ is open, so we can find $g \in C_c(S)$ such that $g \prec E\backslash supp(f)$ and $T(g) > \mu(E\backslash supp(f)) - \varepsilon$. But then $f + g \prec E$, so

$$
\begin{aligned}
\mu(E) \;&>\; T(/) + T(g) > \mu(E \cap U) + \mu(E\backslash supp(f)) - 2\varepsilon \\
&>\; \mu^*(E \cap U) + \mu^*(E\backslash U) - 2\varepsilon.
\end{aligned}
$$

Letting $\varepsilon \to 0$, we obtain the desired inequality.

For the general case, if $\mu^*(E) < \infty$, we can find an open $V \supseteq E$ such that $\mu(V) \le \mu^*(E) + \varepsilon$, and hence

$$
\begin{aligned}
\mu^*(E) + \varepsilon \;&>\; \mu(V) > \mu^*(V \cap U) + \mu^*(V\backslash U) \\
&>\; \mu^*(E \cap U) + \mu^*(E\backslash U).
\end{aligned}
$$

Letting $\varepsilon \to 0$, we are done.

Proof of (III): If K is compact, $f \in C_c(S)$, and $f \ge \chi_K$, let $U_\varepsilon = \{x : f(x) > 1 - \varepsilon\}$. Then U_ε is open, and if $g \prec U_\varepsilon$, we have $g \le (1 - \varepsilon)^{-1}T(f)$ and so $T(g) \le (1 - \varepsilon)^{-1}T(f)$. Thus

$$
\mu(K) \le \mu(U_\varepsilon) \le \frac{1}{1 - \varepsilon}T(f),
$$

and letting $\varepsilon \to 0$ we see that $\mu(K) \le T(/)$. On the other hand, for any open $U \supseteq K$, by **Urysohn's lemma** there exists $f \in C_c(S)$ such that $f \ge \chi_K$ and $f \prec U$, whence $T(f) \le \mu(U)$. Since μ is outer regular on K, **(A2)** follows.

Proof of (IV): If suffices to show that $T(f) = \int f d\mu$ if $f \in C_c(S, [0, 1])$, as $C_c(S)$ is the linear span of the latter set. Given $N \in \mathbb{N}$, for $1 \le j \le N$ let $K_j = \{x : f(x) \ge \frac{j}{N}\}$ and let $K_0 = supp(f)$. Also, define $f_1, ..., f_N \in C_c(S)$ by

$$
\begin{aligned}
f_i(x) \;&=\; 0 \ \text{ if } \ x \notin K_{j-1}, \\
f_j(x) \;&=\; f(x) - \frac{j-1}{N} \ \text{ if } x \in K_{j-1}\backslash K_j, \\
\text{and } \ f_j(x) \;&=\; \frac{1}{N} \ \text{ if } x \in K_j,
\end{aligned}
$$

In other words,

$$
f_j = \min\{\max\{\{f - \frac{j-1}{N}, 0\}, \frac{1}{N}\}.
$$

Then

$$
\frac{1}{N}\chi_{K_j} \le f_j \le \frac{1}{N}\chi_{K_{j-1}},
$$

hence

$$
\frac{1}{N}\mu(K_j) \le \int f_j d\mu \le \frac{1}{N}\mu(K_{j-1}),
$$

Also, $f_j \in C_c(S)$, so if U is an open set containing K_{j-1}, we have $T(f_j) \leq N^{-1}\mu(U)$ and hence, by **(A2)** and outer regularity, Moreover,

$$\frac{1}{N}\mu(K_j) \leq T(f_j) \leq \frac{1}{N}\mu(K_{j-1}).$$

Moreover, $f = \sum_{j=1}^{N} f_j$, so that

$$\frac{1}{N}\sum_{j=1}^{N}\mu(K_j) \;\leq\; \int f d\mu \leq \frac{1}{N}\sum_{j=0}^{N-1}\mu(K_j),$$

$$\frac{1}{N}\sum_{j=1}^{N}\mu(K_j) \;\leq\; T(f) \leq \frac{1}{N}\sum_{j=0}^{N-1}\mu(K_j).$$

It follows that

$$T(f) - \int f d\mu \leq \frac{1}{N}[\mu(K_0) - \mu(K_N)] \leq \frac{1}{N}\mu(supp(f)).$$

Since $\mu(supp(f)) < \infty$ and N is arbitrary, we conclude that $T(f) = \int f d\mu$. $\square$

Another Proof of (IV): **[AB98]**

Let T be a positive linear functional on $C_c(S)$, and let μ be its induced outer measure. Then μ restricted to the Borel sets of S is a regular Borel measure. We shall show that $T(f) = \int f d\mu$ holds for each $f \in C_c(S)$. To this end, let $f \in C_c(S)$. Fix an open set V such that

$$K = Supp(f) \subseteq V \quad \text{and} \quad \mu(V) < \infty.$$

Also, choose $c > 0$ satisfying

$$|f(x)| \leq c \quad \text{for all} \quad x \in S.$$

Given $\varepsilon > 0$, pick N such that $2c/N < \varepsilon$, and let $y_i = -c + i(2c/N)$ for $i = 1, ..., N$; that is, $\{y_0, y_-, ..., y_N\}$ is the partition of $[-c, c]$ with $y_i - y_{i-1} = 2c/N$ for $i = 1, ..., N$. For each $1 \leq i \leq N$, let

$$A_i = \{x \in K : y_{i-1} < f(x) \leq y_i\},$$

and note that the open set

$$W_i = \{x \in V : y_{i-1} - \varepsilon < f(x) \leq y_i + \varepsilon\}$$

satisfies $A_i \subseteq W_i$. Notice that the Borel sets $A_1, ..., A_N$, are pairwise disjoint and $\bigcup_{i=1}^{N} A_i = K$.

By the regularity of μ, for each $1 \leq i \leq N$ there exists an open set V_i satisfying

$$A_i \subseteq V_i \subseteq W_i \quad \text{and} \quad \mu(V_i) - \mu(A_i) < \varepsilon/N.$$

Clearly, $K \subseteq \bigcup_{i=1}^{N} V_i \subseteq V$ holds. By **Partition of Unity**, there exist functions $g_1, ..., g_N \in C_c(S)$ such that $g_i \prec V_i$ for $i = 1, ..., N$ and $\sum_{i=1}^{N} g_i = 1$ for each $x \in K$. Note that $fg_i \leq (y_i + \varepsilon)g_i$ holds for each i and $f = \sum_{i=1}^{N} fg_i$. Therefore,

$$
\begin{aligned}
T(f) - \int f d\mu &= \sum_{i=1}^{N} F(fg_i) - \sum_{i=1}^{N} \int_{A_i} f d\mu \\
&\leq \sum_{i=1}^{N} (y_i + \varepsilon)T(g_i)) - \sum_{i=1}^{N} (y_i - \varepsilon)\mu(A_i) \\
&\leq \sum_{i=1}^{N} (y_i + \varepsilon)\mu(V_i) - \sum_{i=1}^{N} (y_i - \varepsilon)\mu(A_i) \\
&= \sum_{i=1}^{N} (y_i + \varepsilon)[\mu(V_i) - \mu(A_i)] + 2\varepsilon \sum_{i=1}^{N} \mu(A_i) \\
&\leq \sum_{i=1}^{N} (y_i + \varepsilon)\cdot\frac{\varepsilon}{N} + 2\varepsilon\mu(K) \\
&= \varepsilon[c + \varepsilon + 2\mu(K)].
\end{aligned}
$$

Since $\varepsilon > 0$ is arbitrary, $T(f) - \int f d\mu \leq 0$·holds for all $f \in C_c(S)$. Replacing f by $-f$, we get $T(f) - \int f d\mu. \geq 0$. Thus,

$$
T(f) = \int f df\mu \quad \text{for all } f \in C_c(S). \square
$$

Remark: [Fol99, p . 215] The proof of this theorem yields something stronger than the statement: We obtain not just a Borel measure μ but an extension $\widetilde{\mu}$ of μ to the σ-algebra of μ^*-measurable sets. However, it follows from outer regularity that for any $E \subseteq S$,

$$
\mu^*(E) = \inf\{\mu(B) : B \in Bo(S), B \supseteq E\},
$$

so μ^* is the outer measure induced by μ. Further, therefore, $\widetilde{\mu}$ is the completion of μ if μ is $\sigma-$finite and is the saturation of the completion of μ in general.

Remarks (Regularity of Radon measures): [Fol99, p. 209-210] Recall that a Radon measure on S is a Borel measure that is finite on all compact sets, outer regular on all Borel sets, and inner regular on all open sets.

(1) Every Radon measure is inner regular on all of its σ-finite sets. In particular, every σ-finite Radon measure is regular.

(2) If S is σ-compact, every Radon measure on S is regular.

(3) In general, a Radon measure on a non-σ-compact space need not be regular.

(4) Let S be a locally compact Hausdorff space in which every open set is a-compact (which is the case, for example, if S is second countable). Then every Borel measure on S that is finite on compact sets is regular and hence Radon.

The Dual of $(C_0(S), u)$

We recall that, for any locally compact Hausdorff space S, $C_0(S)$ is the uniform closure of $C_c(S)$, and hence if μ is a Radon measure on S, the functional $T(f) = \int f d\mu$ extends continuously to $C_0(S)$ iff it is bounded with respect to the uniform norm. In view of the equality

$$\mu(S) = \sup\{\textstyle\int f d\mu : f \in C_0(S), 0 \le f \le 1\}.$$

(a special case of **(A1)**) together with the fact that $|\int f d\mu| \le \int |f| d\mu$, this happens precisely when $\mu(S) < \infty$, in which case $\mu(S)$ is the operator norm of T.

We have therefore identified the positive bounded linear functionals on $C_0(S)$: they are given by integration against finite Radon measures. Our object in this section is to extend this result to give a complete description of $C_0(S)^*$. The key fact is that real linear functionals on $C_0(S, \mathbb{R})$ have a **"Jordan decomposition."**

Let μ be a real measure and let μ^+ and μ^- be its positive components according to the Hahn-Jordan decomposition. If $f \in L_1(\mu^+) \cap L_1(\mu^-)$, we define the integral of f against the signed measure μ by

$$\textstyle\int_S f \, d\mu = \int_S f \, d\mu^+ - \int_S f \, d\mu^-.$$

The Riesz Representation Theorem 2.1.4. (Riesz-Kakutani) Let S be a locally compact Hausdorff space, and for $\mu \in M(S)$ and $f \in C_0(S)$, let $T_\mu(f) = \int f d\mu$. Then the map $\mu \to T_\mu$ is an isometric isomorphismfrom $M(S)$ to $C_0(S)^*$.

Proof. (Outline) We have already shown that for every $T \in C_0(S)^*$, there are finite Radon measures μ_1 and μ_2 such that $T(f) = \int f d\mu$, where $\mu = \mu_1 - \mu_2$.

On the other hand, if $\mu \in M(S)$, then et

$$|\textstyle\int f d\mu| \le \int |f| d\mu \le ||f||_\infty \cdot ||\mu||,$$

so $T_\mu(f) \in C_0(S)^*$ and $||T_\mu|| \le ||\mu||$. Moreover, if $h = d\mu/d|\mu|$, then $|h| = 1$, so by Lusin's theorem, for any $\varepsilon > 0$, there exists $f \in C_c(S)$ such that $||f||_\infty \le 1$ and $f = \overline{h}$ except on a set E with $|\mu|(E) < \varepsilon/2$. Then

$$\begin{aligned} ||\mu|| &= \textstyle\int |h|^2 d|\mu| = \int \overline{h} d|\mu| \le |\int f d\mu| + |\int (f - \overline{h}) d\mu| \\ &\le |\textstyle\int f d\mu| + 2|\mu|(E) < |\int f d\mu| + \varepsilon.. \end{aligned}$$

It follows that $||\mu|| \le ||T_\mu||$. $\square$

Corollary 2.1.5. If S is a compact Hausdorff space, then $C_0(S)^*$ is isometrically isomorphic to$M(S)$.

Remarks (Baire vs Borel sets): (1) Some authors prefer to restrict attention to a smaller σ-algebra than $Bo(S)$, namely, the σ-algebra $Ba(S)$ generated by $C_c(S)$ (that is, the smallest $\sigma-$algebra with respect to which every $f \in C_c(S)$ is measurable). The elements of $Ba(S)$ are called **Baire sets**. Equivalently, $Ba(S)$ is the σ-algebra generated by the compact $G_\delta-$sets.

(2) Let S be a locally compact Hausdorff space.

(i) If $f \in C_c(S, [0, \infty))$, then $f^{-1}([a, \infty))$ is a compact $G_\delta-$set for all $a > 0$.

(ii) If $K \subseteq S$ is a compact G_δ set, there exists $f \in C_c(S, [0, 1])$ such that $K = f^{-1}(\{1\})$.

(3) Let S be a second countable locally compact Hausdorff space.

(i) Every compact subset of S is a $G_\delta-$set.

(ii). $Bo(S) = Ba(S)$.

(5). Let S be an uncountable set with the discrete topology, or the one-point compactification of such a set. Then $Bo(S) \neq Ba(S)$.

2.2   The Dual of $(C_0(S, X), u)$

Let S be a locally compact Hausdorff space equipped with its Borel $\sigma-$algebra $\mathcal{B}(S)$, and let X be a Banach space. We denote by $C_0(S, X)$ the Banach space (uniform norm) of all continuous functions $f : S \to X$, vanishing at infinity. If $X = \mathbb{R}$, we put $C_0(S, X) = C_0(S)$. According to the Riesz-Kakutani theorem (see [**Rud87, Theorem 6.19**], the dual $C_0^*(S)$ is isometric to the Banach space of all scalar regular measures on S, with the variation norm. All the measures we will deal with here are supposed to be defined on the $\sigma-$algebra $\mathcal{B}(S)$. We denote by X^* the strong dual of X. A theorem of **I. Singer**, settled for S compact, states that the topological dual $C_0^*(S, X)$ is isometrically isomorphic to the Banach space $M_{r\sigma bv}(S, X^*)$ of all regular vector measures of bounded variation on S, with values in the strong dual X^*. Using the **Riesz-Kakutani theorem** and some routine topological arguments, we propose a constructive detailed proof of **Singer's theorem** for S locally compact Hausdorff space, which is, as far as we know, different from that supplied elsewhere [**Sing57; Hen96**].

Recall the definition of the Banach space $M_{r\sigma bv}(\mathcal{B}(S), X^*)$ given in **Definition 1.4.3**, and **Theorem 1.4.4**.

Theorem 2.2.1. *There is an isometric isomorphism between the topological dual $C_0^*(S, X)$ of $C_0(S, X)$ and the Banach space $M_{r\sigma bv}(\mathcal{B}(S), X^*)$, where the functional $U \in C_0^*(S, X)$ and the corresponding measure $\lambda \in M_{r\sigma bv}(\mathcal{B}(S), X^*)$ are related by the integral formula:*

$$Uf = \int_S f.d\lambda, \quad \|U\| = |\lambda|, f \in C_0(S, X), \qquad (\#2.1)$$

*where the integral is the one defined in **Section 1.1.5**.*

Remark. Actually the original proof of this theorem [**Sing57**] contains some gaps about the strong $\sigma-$additivity and regularity of the measure λ attached to the functional U. These gaps has been filled by **J. Gil de Lamadrid** in [**Lam66**]. Another proof using the **Hahn-Banach theorem** and measures on product spaces. can be found in Hensgen [**Hen96**]. To settle the proof of the theorem we present some preparatory lemmas. Let us start with a $U \in C_0^*(S, X)$, we will construct a $\lambda \in M_{r\sigma bv}(\mathcal{B}(S), X^*)$ such that formula (**#2.1**) holds.

Lemma 2.2.2. *For each $f \in C_0(S)$ denote by $W(f)$ the functional on X given by:*

$$W(f)(x) = U(f.x), \qquad x \in X, \qquad\qquad (\#2.2)$$

Then W is a linear bounded operator from $C_0(S)$ into X^* and $\|W\| \leq \|U\|$.

Proof: It is clear that W is linear, we show that $W(f) \in X^*$. Indeed $|W(f)(x)| = |U(f.x)| \leq \|U\| \cdot \|f\|_\infty \cdot \|x\|$, so $\|W(f)\| \leq \|U\| \cdot \|f\|_\infty$ thus

$$W(f) \in X^* \text{ and } \|W\| \leq \|U\|. \ \square$$

Lemma 2.2.3. *For each fixed $x \in X$, there exists a unique scalar regular measure μ_x on $\mathcal{B}(S)$, such that:*

$$W(f)(x) = \int_S f d\mu_x, \ f \in C_0(S) \qquad\qquad (\#2.3)$$

Proof: Define $W_x : C_0(S) \to \mathbb{R}$ by

$$W_x(f) = W(f)(x), f \in C_0(S),$$

then W_x is linear and bounded that is $W_x \in C_0^*(S)$ and we have

$$|W_x(f)| \leq \|U\| \cdot \|f\|_\infty \cdot \|x\|,$$

therefore

$$\|W_x| \leq \|U\| \cdot \|x\|.$$

By the Riesz-Kakutani theorem there is a unique scalar regular measure μ_x on $\mathcal{B}(S)$ such that:

$$W_x(f) = W(f)(x) = \int_S f d\mu_x, \ \forall f \in C_0(S) \text{ and } \|W_x\| = |\mu_x|.$$

$\square$

Lemma 2.2.4. *Define the set function λ on $\mathcal{B}(S)$ by the following recipe: for $A \in \mathcal{B}(S)$, $\lambda(A)$ is the functional on X given by:*

$$x \in X, \ \lambda(A)x = \mu_x(A), x \in X \qquad\qquad (\#2.4)$$

where μ_x comes from **Lemma 2.2.3**. *Then $\lambda(A) \in X^*$ for each $A \in \mathcal{B}(S)$, moreover λ is additive.*

Proof: Let x, $y \in X$, $A \in \mathcal{B}(S)$, then $\lambda(A)\,(x+y) = \mu_{x+y}(A)$ where μ_{x+y} corresponds to W_{x+y} according to (**#2.4**), thus

$$W_{x+y}\,(f) = \int_S f d\mu_{x+y}, \ \forall f \in C_0\,(S)\,.$$

Since $W_{x+y}\,(f) = W\,(f)\,(x+y) = W\,(f)\,(x) + W\,(f)\,y$, we deduce from (**#2.4**) that:

$$W_{x+y}\,(f) = \int_S f d\mu_{x+y} = \int_S f d\mu_x + \int_S f d\mu_y = \int_S f d\left(\mu_x + \mu_y\right),$$

where the last equality is easy to check by standard method.

Thus

$$\int_S f d\mu_{x+y} = \int_S f d\left(\mu_x + \mu_y\right), \ \forall f \in C_0\,(S)\,.$$

From the fact that $\mu_x + \mu_y$ is regular, the uniqueness part of the **Riesz-Kakutani theorem** yields $\mu_{x+y} = \mu_x + \mu_y$. Likewise $\mu_{\alpha x} = \alpha \mu_x$, for $\alpha \in \mathbb{R}$. This proves that $\lambda(A)$ is a linear functional on X. On the other hand we have:

$$|\lambda(A)x| = |\mu_x(A)| \leq |\mu_x|\,(A) \leq \|\mu_x\| = \|W_x\| \leq \|U\| \cdot \|x\|$$

(see the proof of **Lemma 2.2.3**). So we deduce that $\lambda(A) \in X^*$ and

$$\|\lambda(A)\| \leq \|U\| \ \forall A \in \mathcal{B}(S).$$

Finally, it is clear that λ is additive. $\square$

The remaining lemmas are intended to prove that the additive set function λ is actually a vector measure. The following is crucial:

Lemma 2.2.5. *The set function λ has finite variation. Moreover, we have $|\lambda| \leq \|U\|$.*

Proof: We use the formula given in **Theorem 1.4.4** for the variation of λ. Let A_1, $A_2,\ldots,A_n$ be a finite partition of the compact space S by sets in $\mathcal{B}(S)$ and let x_1, $x_2,\ldots,x_n$ be vectors in X with $\|x_i\| \leq 1$ $\forall i$. We need an estimation of the sum $\sum_{1}^{n} \lambda(A_i)x_i$. Let $\varepsilon > 0$, then by the regularity of the measures μ_{x_i}, there exist compact sets K_1, $K_2,\ldots,K_n$ and open sets G_1, $G_2,\ldots,G_n$ such that: $K_i \subseteq A_i \subseteq G_i$ and

$$\left|\mu_{x_i}\right|(G_i \backslash K_i) < \frac{\varepsilon}{2n}, i = 1, 2, \ldots n.$$

Note that the K_i are pairwise disjoint since A_i are so. Since S is locally compact, disjoint compact sets have disjoint neighbourhoods (note, this is not a consequence of normality, because a locally compact, not compact space, may be not normal). So, using a simple induction on n, we can construct pairwise disjoint open sets U_1, $U_2,\ldots,U_n$ such that $K_i \subseteq U_i$ $\forall i$, letting $V_i = U_i \cap G_i$, we get pairwise disjoint open sets V_i such that $K_i \subseteq V_i \subseteq G_i$ $\forall i$.

Now, Let $g_i : S \to \mathbb{R}$ be a continuous function such that $0 \le g_i(t) \le 1$ $\forall t \in S$, $g_i(t) = 1$ $\forall t \in K_i$, support $g_i \subseteq V_i$ (such functions exist by Urysohn's lemma since S is locally compact). We have:

$$\int_S g_i d\mu_{x_i} = \int_{V_i} g_i d\mu_{x_i}$$

(since $g_i \equiv 0$ outside V_i), so we deduce that:

$$\int_S g_i d\mu_{x_i} = \int_{V_i \setminus K_i} g_i d\mu_{x_i} + \int_{K_i} g_i d\mu_{x_i}.$$

But $\int_{K_i} g_i \, d\mu_{x_i} = \mu_{x_i}(K_i)$ (because $g_i \equiv 1$ on K_i). consequently we have

$$\int_S g_i d\mu_{x_i} - \mu_{x_i}(K_i) = \int_{V_i \setminus K_i} g_i d\mu_{x_i}.$$

This gives the following estimation:

$$\begin{aligned}
\left| \int_S g_i d\mu_{x_i} - \mu_{x_i}(K_i) \right| &= \left| \int_{V_i \setminus K_i} g_i d\mu_{x_i} \right| \le \int_{V_i \setminus K_i} g_i d. \left| \mu_{x_i} \right| \\
&\le \left| \mu_{x_i} \right| (V_i \setminus K_i) \quad (\text{since } 0 \le g_i \le 1) \\
&\le \left| \mu_{x_i} \right| (G_i \setminus K_i) \quad (\text{since } V_i \subseteq G_i) \\
&< \frac{\varepsilon}{2n}
\end{aligned}$$

Therefore

$$\left| \int_S g_i d\mu_{x_i} - \mu_{x_i}(K_i) \right| < \frac{\varepsilon}{2n}, \ \forall i, \tag{\#2.5}$$

Now, let $f : S \to X$ be the function defined by:

$$f(t) = \sum_{i}^{n} g_i(t) . x_i, t \in S.$$

then f is continuous and we have: $f(t) = 0$ for each t in $S \setminus \bigcup_{1}^{n} V_i$, and $f(t) = g_i(t) . x_i$ for each t in V_i, because V_i are pairwise disjoint and support $g_i \subseteq V_i$. Then we deduce that $\|f\| \le 1$ and

$$Uf = \sum_{1}^{n} U(g_i . x_i) = \sum_{1}^{n} \int_S g_i d\mu_{x_i},$$

by (#2.5) since $U(g_i . x_i) = W(g_i)(x_i)$. So

$$\begin{aligned}
\left| Uf - \sum_{1}^{n} \mu_{x_i}(K_i) \right| &= \left| \sum_{1}^{n} \int_S g_i d\mu_{x_i} - \sum_{1}^{n} \mu_{x_i}(K_i) \right| \\
&\le \sum_{1}^{n} \left| \int_S g_i d\mu_{x_i} - \mu_{x_i}(K_i) \right| \\
&< \sum_{1}^{n} \frac{\varepsilon}{2n} \text{ by (5)} = \frac{\varepsilon}{2}.
\end{aligned}$$

Therefore

$$\left| Uf - \sum_1^n \mu_{x_i}(K_i) \right| < \frac{\varepsilon}{2}. \qquad (\#2.6)$$

Now, we turn to the estimation of $\left| \sum_1^n \lambda(A_i)x_i \right|$.

$$\left| \sum_1^n \lambda(A_i)x_i \right| - |Uf| \leq \left| \sum_1^n \lambda(A_i)x_i - Uf \right|$$

$$\leq \left| \sum_1^n \lambda(A_i)x_i - \sum_1^n \mu_{x_i}(K_i) \right| + \left| Uf - \sum_1^n \mu_{x_i}(K_i) \right|$$

and

$$\left| \sum_1^n \lambda(A_i)x_i - \sum_1^n \mu_{x_i}(K_i) \right| = \left| \sum_1^n \mu_{x_i}(A_i) - \sum_1^n \mu_{x_i}(K_i) \right|$$

$$\leq \sum_1^n |\mu_{x_i}|(A_i \backslash K_i) \leq \sum_1^n |\mu_{x_i}|(G_i \backslash K_i)$$

$$< \sum_1^n \frac{\varepsilon}{2n} = \frac{\varepsilon}{2}$$

Combining this with (6) we get

$$\left| \sum_1^n \lambda(A_i)x_i \right| - |Uf| < \frac{\varepsilon}{2} + \frac{\varepsilon}{2} = \varepsilon,$$

so

$$\left| \sum_1^n \lambda(A_i)x_i \right| < |Uf| + \varepsilon \leq \|U\| \cdot \|f\|_\infty + \varepsilon$$

$$\leq \|U\| + \varepsilon \quad (\text{since } \|f\| \leq 1)$$

Letting $\varepsilon \searrow 0$ we obtain

$$\left| \sum_1^n \lambda(A_i)x_i \right| \leq \|U\|.$$

This estimation is valid for all finite partitions $\{A_i\}$ of S in $\mathcal{B}(S)$ and all systems $\{x_i\}$ in X with $\|x_i\| \leq 1$. So, by taking the appropriate supremum, this leads to $|\lambda|(S) \leq \|U\| < \infty$, by **Theorem 1.4.4**. Then λ has finite variation. $\square$

Lemma 2.2.6. *For each $A \in \mathcal{B}(S)$ we have:*

$$|\lambda|(A) = \sup\{|\lambda|(K) : K \subseteq A, K \text{ compact}\}, \qquad (\#2.7)$$
$$|\lambda|(A) = \sup\{|\lambda|(G) : A \subseteq G, G \text{ open}\}, \qquad (\#2.8)$$

In other words the variation measure $|\lambda|$ of λ is regular, and so λ is regular.

Proof: Let $A \in \mathcal{B}(S)$, since $|\lambda| < \infty$, (7) is equivalent to the following approximation: For each $\varepsilon > 0$, there is a compact K such that:

$$K \subseteq A, \ |\lambda|(A) - \varepsilon < |\lambda|(K). \qquad (\#2.9)$$

Let $\varepsilon > 0$. Again since $|\lambda| < \infty$ there exists a finite partition E_1, $E_2, \ldots, E_n$ of A in $\mathcal{B}(S)$ and $x_1, x_2, \ldots, x_n$ in X with $\|x_i\| \leq 1 \ \forall i$, such that

$$|\lambda|(A) - \frac{\varepsilon}{2} < \left| \sum_1^n \lambda(E_i) x_i \right|,$$

by formula in **Theorem 1.4.4**.

By formula $(\#2.4)$, the measures $\lambda(\bullet) x_i = \mu_{x_i}(\bullet)$ are regular; consequently, there exist compact sets $K_1, K_2, \ldots, K_n$, with $K_i \subseteq E_i$ and

$$|\lambda(E_i \backslash K_i) x_i| < \frac{\varepsilon}{2n} \ \forall i.$$

Then we have:

$$|\lambda|(A) - \frac{\varepsilon}{2} \ < \ \left| \sum_1^n \lambda(E_i) x_i \right| \leq \left| \sum_1^n \lambda(K_i) x_i \right| + \left| \sum_1^n \lambda(E_i \backslash K_i) x_i \right|$$

$$\leq \ \sum_1^n |\lambda(K_i) x_i| + \sum_1^n |\lambda(E_i \backslash K_i) x_i| < |\lambda|(K) + \frac{\varepsilon}{2},$$

where K is defined to be the compact set $\bigcup_{i=1}^n K_i$. Therefore, $(\#2.9)$ is valid and proves $(\#2.7)$.

We can get $(\#2.8)$ by applying $(\#2.7)$ to the complement A^c of the set A. $\square$

Lemma 2.2.7. *The variation measure $|\lambda|$ is σ-additive.*

Proof: Since λ is additive then so is $|\lambda|$. By the regularity property just proved, the result is a consequence of **Alexandroff theorem** (see **[DS58, p.138]**). $\square$

Lemma 2.2.8. *The set function λ is a regular vector measure, that is λ is a member of $M_{r\sigma bv}(\mathcal{S}, X^*)$.*

Proof: We know that λ is additive, so to prove the σ-additivity it is enough to prove the continuity at $\varnothing$, that is for every sequence A_n in $\mathcal{B}(S)$ decreasing to $\varnothing$, we have $\lambda(A_n) \to 0$. But it is a consequence of the σ-additivity of $|\lambda|$ and the fact that

$$\|\lambda(A)\| \leq |\lambda|(A), \forall A \in \mathcal{B}(S).$$

On the other hand λ is regular since $|\lambda|$ is regular by **Lemma 2.2.6**. $\square$

Lemma 2.2.9. Let $v, \mu \in M_{r\sigma bv}(\mathcal{B}(S), X^*)$ be such that

$$\int_S f \, dv = \int_S f \, d\mu \text{ for all } f \in C_0(S, X).$$

Then $v \equiv \mu$.

Proof: Take $f \in C_0(S, X)$ of the form $f = g \otimes x$, where $g \in C_0(S)$ and x fixed in X. Then by standard tools we have

$$\int_S f dv = \int_S g dv\,(\bullet)\,x$$

and

$$\int_S f d\mu = \int_S g d\mu\,(\bullet)\,x.$$

This yields

$$\int_S g dv\,(\bullet)\,x = \int_S g d\mu\,(\bullet)\,x.$$

Since both scalar measures $v\,(\bullet)\,x$ and $\mu\,(\bullet)\,x$ are regular and since g is arbitrary, we deduce from Riesz-Kakutani theorem that $v\,(\bullet)\,x = \mu\,(\bullet)\,x$ for each $x \in X$. Thus $v \equiv \mu$. $\square$

Now, we are in a position to give the proof of **Theorem 2.2.1**.

Proof of Theorem 2.2.1: First we prove relation (**#2.1**), i.e,

$$Uf = \int_S f d\lambda, \text{ for all } f \in C_0(S, X),$$

where λ is the vector measure constructed in **Lemma 2.2.4**.

Let $f \in C_0(S, X)$ be of the form $f = g \otimes x$, where $g \in C_0(S)$ and x fixed in X. Then

$$
\begin{aligned}
Uf \;&=\; W\,(g)\,(x) = W_x\,(g) = \int_S g d\mu_x \ \text{ by (\#2.3)} \\
&=\; \int_S g d\lambda\,(\bullet)\,x \ \text{ by (\#2.4)}.
\end{aligned}
$$

But we have

$$\int_S g d\lambda\,(\bullet)\,x = \int_S g.x.d\lambda.$$

Therefore, formula (**3**) is satisfied for $f = g \otimes x$. By linearity we can see that formula (**#2.1**) is satisfied for all $f \in C_0(S) \otimes X$, the vector space of all $f \in C_0(S, X)$ of the form

$$f = \sum_1^n g_i \otimes x_i, \text{ with } g_i \in C_0(S),\ x_i \in X,\ i = 1, 2..., n.$$

It is well known that $C_0(S) \otimes X$ is dense in $C_0(S, X)$ (see **Section A.5**, or [**Din67, Proposition 1, Section 19**]). Consequently, if $f \in C_0(S, X)$, there is a sequence f_n in $C_0(S) \otimes X$ converging to f uniformly on S. By the integration process with respect to an operator valued measure we get:

$$\left| \int_S f_n\, d\lambda - \int_S f\, d\lambda \right| \leq \|f_n - f\|_\infty .\tilde{\lambda}\,(S),$$

where $\tilde{\lambda}$ is the semivariation of λ defined by the RHS of formula in **Definition 1.4.2** and which is, in the present context, equal to the variation $|\lambda|$ (see **Theorem 1.4.4**). As λ is of finite variation and $\|f_n - f\|_\infty \to 0$, we have

$$\int_S f_n d\lambda \to \int_S f d\lambda.$$

But, $Uf_n = \int_S f_n \, d\lambda$ because $f_n \in C_0(S) \otimes X$ $\forall n$. Since U is bounded and $f_n \to f$ uniformly we get

$$Uf_n = \int_S f_n d\lambda \to Uf.$$

Hence

$$Uf = \int_S f d\lambda \ \forall f \in C_0(S, X).$$

By **Lemma 2.2.9** λ is the unique measure in $M_{r\sigma bv}(\mathcal{B}(S), X^*)$ satisfying relation **(1)**. This proves that the correspondence $U \xrightarrow{\varphi} \lambda$ from $C_0^*(S, X)$ into $M_{r\sigma bv}(\mathcal{B}(S), X^*)$ is well-defined. Moreover we have

$$|Uf| = \left|\int_S f d\lambda\right| \le \|f\|_\infty \cdot \tilde{\lambda}(S) = \|f\|_\infty \cdot \|\lambda\|,$$

so $\|U\| \le |\lambda|$ and by **Lemma 2.2.5** we get $\|U\| = |\lambda|$. This implies that φ is an isometry and then it is 1-1. It is not difficult to show that φ is linear. To complete the proof, we must show that φ is onto. To this end, let us start with $\mu \in M_{r\sigma bv}(S, X^*)$, to which we associate the functional on $C_0(S, X)$ given by:

$$Uf = \int_S f d\mu, f \in C_0(S, X).$$

It is clear that U is linear and bounded, so $U \in C_0^*(S, X)$. We show that $\varphi(U) = \mu$. Put $\varphi(U) = \lambda$, that is λ is the vector measure constructed along **Lemmas 2.2.2 - 2.2.4**. Then by formula **(1)**,

$$Uf = \int_S f d\lambda \ \forall f \in C_0(S, X),$$

which yields

$$\int_S f d\mu = \int_S f d\lambda \ \forall f \in C_0(S, X).$$

From **Lemma 2.2.9**, we deduce that $\mu = \lambda$, and this completes the proof of theorem. $\square$

2.3   The Dual of $(C_b(S), \beta)$

The following version of the classical Riesz representation theorem is well-known (see [**Rud87, Roy88, DS58**]).

Theorem 2.3.1 (**Riesz-Markov**, 1938) *Let S be a compact Hausdorff space. Then the u-dual of $C(S)$ is isometrically isomorphic to $M(S)$. More precisely, for any $\mu \in M(S)$, the equation*

$$L(f) = \int_S f d\mu \qquad (f \in C(S)) \tag{#2.10}$$

defines a u-continuous linear functional L on $C(S)$ such that $\|L\| = \|\mu\|$. Conversely, for any u-continuous linear functional L on $C(S)$, there exists *a unique $\mu \in M(S)$ such that L is given by (#2.10) and $||L|| = ||\mu||$. In short, we write $(C_b(S), u)^* \cong M(S)$*

Theorem 2.3.2. (Riesz-Alexandroff) Let S be a completely regular Hausdorff space. and βS the Stone-Cech compactification of S. Then $(C_b(S), u)^* \cong M(\beta S)$.

Proof. This follows from the fact that $(C_b(S), u)$ is isometrically isomorphic to $(C_b(\beta S), u)$; hence by above theorem, $(C_b(S), u)^* \cong (C_b(\beta S), u)^* \cong M(\beta S)$.

Theorem 2.3.3. (Riesz-Alexandroff) Let S be a completely regular Hausdorff space. Then $(C_b(S), u)^* \cong M(\beta S)$ *via the linear isomorphism* $L \longleftrightarrow \mu$, *where*

$$L(f) = \int_S f d\mu \text{ for all } f \in C_b(S) \text{ and } ||L|| = |\mu|(S).$$

Proof. This follows from the fact that $(C_b(S), u)$ is isometrically isomorphic to $(C_b(\beta S), u)$; hence by above theorem, $(C_b(S), u)^* \cong (C_b(\beta S), u)^* \cong M(\beta S)$.

Theorem 2.3.4. **(Riesz-Markov)** Let S be a locally compact Hausdorff space. Then $(C_0(S), u)^* \cong M(S)$

Proof. See, [**Rud87, Theorem 6.19**].

Recall that the strong topology on $(C_b(S), \beta)^*$ is, by definition, the topology of uniform convergence on β-bounded subsets of $C_b(S)$. Since the β-bounded and the u-bounded subsets of $C_b(S)$ are the same, the strong topology on $(C_b(S), \beta)^*$ coincides with that of the norm topology of $(C_b(S), u)^*$.

The following result, established independently by Giles [**Gil72**], Gulick [**Gul72**], Hoffman-Jørgensen [**HJ72**], extends Buck's theorem to a completely regular Hausdorff space S.

Theorem 2.3.5. **(Riesz-Buck-Giles)** *Let S be a completely regular Hausdorff space. Then the β-dual of $C_b(S)$, with the strong topology is isometrically isomorphic to $M(S)$.*

Proof. [**Gil72**] Let $\mu \in M(S)$. We first show that the linear functional L on $C_b(S)$ defined by the equation

$$L(f) = \int_S f d\mu, \ f \in C_b(S),$$

is β-continuous. Without loss of generality, we may assume that $\mu \geq 0$.and $\mu(S) = 1$. The integral on the R.H.S. exists and the linearilty of L follows by standard properties of the integral.

Since μ is regular, there exists a sequence $\{K_n : n = 0, 1, 2, ...\}$ of compact sets in S such that $\varnothing = K_0, K_n \subseteq K_{n+1}$, and $\mu(S \backslash K_n) \leq \frac{1}{2^{2n}}$. Let

$$\varphi = \sum_{n=0}^{\infty} \frac{1}{2^{n-1}} \chi_{K_n}.$$

Then $\varphi \in B_0(S)$, and, for each $x \in K_{n+1} \backslash K_n$,

$$\frac{1}{2^n} < \varphi(x) = \frac{1}{2^{n-1}}.$$

Since φ is upper-semicontinuous, it is μ-measurable function. Moreover,

$$\varphi = 0 \ \text{ outside } \ Y = \bigcup_{n=1}^{\infty} K_n, \text{ and } \ \mu(S\backslash Y) = \lim_{n\to\infty} \mu(S\backslash K_n) = 0.$$

Therefore

$$\int_S \frac{1}{\varphi} d\mu = \int_Y \frac{1}{\varphi} d\mu = \sum_{n=0}^{\infty} 2^{n-1}\mu(K_{n+1}\backslash K_n) \leq 1.$$

If $f \in U(\varphi,1) = \{f \in C_b(S) : \|\varphi f\| \leq 1\}$, then

$$|L(f)| = |\int_S f d\mu \ \leq \|\varphi f\|.\int_S \frac{1}{\varphi} d\mu \leq 1,$$

and so β-continuous.

Conversely, Let L be a β-continuous linear functional on $C_b(S)$. Since $\beta \leq u$, L be a u-continuous linear functional on $C_b(S)$ and so by the **Riesz-Alexandroff Theorem**, there exists a unique $\widehat{\mu} \in M(\beta S)$ such that

$$L(f) = \int_S \widehat{f} d\widehat{u} \text{ for all } \ f \in C_b(S),$$

where $\widehat{f}$ is the unique extension of f to βS. We now show that $\widehat{\mu}$ has σ-compact support in S.

Suppose this is not true. Then there exists an $\varepsilon > 0$ such that

$$\widehat{\mu}(\beta S\backslash K) > \varepsilon \text{ for every compact } K \subseteq S. \tag{\#2.11}$$

Since L be a β-continuous linear functional on $C_b(S)$, there exists a $\varphi \in B_0(S)$ such that

$$|L(f)| < 1 \text{ for all } f \in U(\varphi,1). \tag{\#2.12}$$

Let $K_1 \subseteq S$ be a compact set such that

$$|\varphi(x)| < \frac{\varepsilon}{4} \ \text{ if } x \notin K_1. \tag{\#2.13}$$

Since $\widehat{\mu}$ is regular and $\beta S\backslash K_1$ is open in βS, we can choose by $(\#2.11)$ a compact set $\widehat{F} \subseteq \beta S\backslash K_1$ such that

$$|\widehat{\mu}|(\widehat{F}) > \frac{\varepsilon}{2}. \tag{\#2.14}$$

Moreover, there exists an open set $\widehat{G}$ such that

$$\widehat{F} \subseteq \widehat{G} \subseteq \beta S\backslash K_1 \ \text{ and } \ |\widehat{\mu}|(\widehat{G}\backslash\widehat{F}) < \frac{\varepsilon}{4}. \tag{\#2.15}$$

Let $\widehat{f}$ be a function in $C(\beta S)$ such that

$$0 \leq \widehat{f} \leq \frac{4}{\varepsilon}, \widehat{f} = \frac{4}{\varepsilon} \ \text{ on } \ \widehat{f} = 0 \text{ on } \beta S\backslash\widehat{G}. \tag{\#2.16}$$

In particular, $\widehat{f} = 0$ on K_1. Define $f = \widehat{f}|S$. Then $f \in C_b(S)$ and, by using (#2.13) and (#2.16),

$$\|\varphi f\| = \|\varphi f\|_{S \setminus K_1} < \frac{\varepsilon}{4} \cdot \frac{4}{\varepsilon} = 1.$$

But using (#2.14)-(#2.1.6)

$$
\begin{aligned}
|L(f)| &= |\int_{\beta S} \widehat{f} d\widehat{\mu}| = |\int_{\widehat{G}} \widehat{f} d\widehat{\mu}| = |\int_{\widehat{G}} \widehat{f} d\widehat{\mu} + \int_{\widehat{G} \setminus \widehat{F}} \widehat{f} d\widehat{\mu}| \\
&\geq |\int_{\widehat{F}} \widehat{f} d\widehat{\mu}| - |\int_{\widehat{G} \setminus \widehat{F}} \widehat{f} d\widehat{\mu}| \geq \frac{4}{\varepsilon} |\int_{\widehat{F}} d\widehat{\mu}| - \frac{4}{\varepsilon} |\int_{\widehat{G} \setminus \widehat{F}} d\widehat{\mu}| \\
&= \frac{4}{\varepsilon} |\widehat{\mu}|(\widehat{F}) - \frac{4}{\varepsilon} |\widehat{\mu}|(\widehat{G} \setminus \widehat{F}) > \frac{4}{\varepsilon} \cdot \frac{\varepsilon}{2} - \frac{4}{\varepsilon} \cdot \frac{\varepsilon}{4} = 1,
\end{aligned}
$$

which contradicts (#2.12). Thus there is a σ-compact set $D \subseteq S$ such that $|\widehat{\mu}|(\beta S \setminus D) = 0$ (so in particular, $|\widehat{\mu}|(\beta S \setminus S) = 0$. It now follows that S, and hence every Borel set in S, is $\widehat{\mu}$-measurable.

Let $\mu(B) = \widehat{\mu}(B)$ for every $B \in Bo(S)$. Then $\mu \in M(S)$ and

$$L(f) = \int_{\beta S} \widehat{f} d\widehat{\mu} = \int_S f d\widehat{\mu} \qquad (f \in C_b(S)).$$

To prove the isometry part, note that the strong topology on $(C_b(S), \beta)^*$ is, by definition, the topology of uniform convergence on β-bounded subsets of $C_b(S)$. Since the β-bounded and the u-bounded subsets of $C_b(S)$ are the same, the strong topology on $(C_b(S), \beta)^*$ coincides with that of the norm topology of $(C_b(S), u)^*$. Thus

$$\|L\| = \sup_{f \in C_b(S), \|f\| \leq 1} |L(f)| = \sup_{f \in C_b(S), \|f\| \leq 1} |\int_S f d\mu| = |\mu|(S) = \|\mu\|.$$

2.4 The Dual of $(C_b(S, X), \beta)$, X Locally Convex

Let S be a completely regular Hausdorff space and X a real LCS with dual X^* and whose topology is generated by a family $cs(X)$ of continuous seminorms. Let $C_{pc}(S, X)$ (resp. $C_{rc}(S, X)$) denote the subspace of $C_b(S, X)$ consisting of those functions f such that $f(S)$ is precompact (resp. relatively compact). Note that if X is complete, then $C_{pc}(S, X) = C_{rc}(S, X)$.

In this section, we first define and establish existence of the integral of functions in $C_b(S, X)$ with respect to certain X^*-valued measures, and then obtain Riesz representation type theorems for the characterization of dual space of $(C_{pc}(S, X), \beta)^*$.

Integral of a vector valued function against a Dual Space Valued Measure

We shall use here the measure theoretic terminology. Let $z(S)$ (resp. $c(S)$) be the collection of all zero (resp. closed) subsets of S. Let $\mathcal{B}(S)$ denote the algebra generated by $z(S)$ and let $\mathcal{B}a(S)$ (resp. $\mathcal{B}o(S)$) denote the σ-algebra generated by $z(S)$(resp. $c(S)$). The space of all Baire measures is denoted by $M(\mathcal{B}a(S))$. Let $M_t(\mathcal{B}o(S))$ denote the space of all bounded real-valued countably additive tight measures on $\mathcal{B}o(S)$.

Definition 2.4.1. For each $p \in cs(X)$, let $M_p(\mathcal{B}(S), X^*)$ denote the set of all finitely-additive X^*-valued set functions m on $\mathcal{B}(S)$ such that

(i) for each $a \neq 0$ in X, $m_a(F) = m(F)(a)$ $(F \in \mathcal{B}(S))$ determines an element m_a of $M(\mathcal{B}(S))$;

(ii) there exists a constant $\lambda > 0$ such that $|m|_p(S) \leq \lambda$, where, for each $F \in \mathcal{B}(S)$, we define $|m|_p$ by

$$|m|_p(F) = \sup |\sum_i m(F_i)(a_i)|,$$

(it would be the same for a balanced W) the supremum being taken over all finite partitions $\{F_i\}$ of F into sets in $\mathcal{B}(S)$ (henceforth referred to as a $\mathcal{B}(S)$-partition) and all finite collections $\{a_i\} \subseteq W$.

Let

$$M(\mathcal{B}(S), X^*) = \bigcup_{p \in cs(X)} M_p(\mathcal{B}(S), X^*).$$

Similarly, we may define the space $M_t(\mathcal{B}o(S), X^*)$ by replacing $M(\mathcal{B}(S))$ by $M_t(\mathcal{B}o(S))$, respectively, in the above definition.

Lemma 2.4.2. [KR81, Kat83] *Let* $p \in cs(X)$. *If* $m \in M_p(\mathcal{B}(S), X^*)$ *(resp.* $M_{t,p}(\mathcal{B}o(S), X^*))$, *then* $|m|_p \in M(\mathcal{B}(S))$ *(resp.* $(M_t(\mathcal{B}o(S))))$.

Proof. First, let $m \in M_p(\mathcal{B}(S), X^*)$. It follows immediately from the definition that $|m|_p$ is a bounded non-negative set function on S, and it is straightforward to show that $|m|_p$ is finitely additive. To show that $|m|_p$ is regular, let $F \in \mathcal{B}(S)$ and $\varepsilon > 0$. There exist a $\mathcal{B}(S)$-partition $\{F_1, ..., F_n\}$ of F and a collection $\{a_1, ..., a_n\} \subseteq W = \{a \in X : p(a) \leq 1\}$ such that

$$|m|_p(F) \leq \sum_{i=1}^{n} |m_{a_i}|(F_j) + \varepsilon$$

Since each m_{a_i} is regular, there exist zero sets $Z_i(i = 1, ..., n)$ such that $Z_i \subseteq F_i$ and

$$|m_{a_i}|(F_j) < |m_{a_i}|(Z_j) + \varepsilon/2^i.$$

Let $Z = \bigcup_{i=1}^{n} Z_i$. Then $Z \subseteq F$ and

$$|m|_p(F) \leq \sum_{i=1}^{n} |m_{c_i}|(Z_i) + 2\varepsilon \leq |m|_p(F) + 2\varepsilon.$$

Thus $|m|_p \in M(S)$.

Now, let $m \in M_{t,p}(\mathcal{B}o(S), X^*)$. Then, by the above argument, it easily follows that $|m|_p$ is a bounded, non-negative, finitely-additive set function on $\mathcal{B}o(S)$. We show that $|m|_p$ is countably additive, as follows. Let $\{A_k : k = 1, 2, ...\}$ be a sequence of disjoint sets in $\mathcal{B}o(S)$ and suppose that $\bigcup_{k=1}^{\infty} A_k = A$. For any $n \geq 1$,

$$|m|_p (A) \geq |m|_p \left(\bigcup_{k=1}^{n} A_k \right) = \sum_{k=1}^{n} |m|_p (A_k),$$

and so

$$|m|_p (A) \geq \sum_{k=1}^{\infty} |m|_p (A_k).$$

Let $\varepsilon > 0$. Then there exist a $\mathcal{B}o(S)$-partition $\{F_j : j = 1, ..., q\}$ of A and a collection of points $\{a_j : j = 1, ..., q\} \subseteq W$ such that

$$|m|_p (A) \leq \sum_{j=1}^{q} m_{a_j} (F_j) + \varepsilon.$$

Since each m_{a_j} is countably additive and $\{F_j \cap A_k : k = 1, 2, ...\}$ is a partition of F_j, we have $m_{a_j}(F_j) = \sum_{k=1}^{\infty} m_{a_j}(F_j \cap A_k)$. Hence

$$|m|_p (A) \leq \left| \sum_{j=1}^{q} \sum_{k=1}^{\infty} m_{a_j} (F_j \cap A_k) \right| + \varepsilon \leq \sum_{k=1}^{\infty} |m|_p (A) + \varepsilon.$$

Since ε is arbitrary, it follows from the above inequalities that $|m|_p$ is countably additive. Next, using the argument of the first part, it easily follows that $|m|_p$ is $k(W)$-regular. Thus $|m|_p \in M_t(\mathcal{B}o(S))$. $\square$

Integration of Vector-valued Functions

Definition 2.4.3.. [**Wel65, Kat74**] If $f \in C_b(S, X)$ and $p \in cs(X)$, we write

$$\|f\|_p = \|p \circ f\| = \sup_{x \in S} |p(f(x))| \,.$$

Let $m \in M_p(\mathcal{B}(S), X^*)$, and $f \in C_b(S, X)$. Let D be the collection of all $\alpha = \{F_1, ..., F_n; x_1, ..., x_n\}$, where $\{F_i\}$ ($1 \leq i \leq n$) is a $\mathcal{B}(S)$-partition of S and $x_i \in F_i$. If $\alpha_1, \alpha_2 \in D$, define $\alpha_1 \geq \alpha_2$ iff each set which appears in α_1 is contained in some set in α_2. In this way, D becomes as indexing set. Let $S_\alpha = \sum_{i=1}^{n} m(F_i)(f(x_i))$. We now define the *integral* of f with respect to m over S by

$$\int_S f dm = \lim_{\alpha \in D} S_\alpha.$$

Regarding the conditions under which this integral exists, we obtain the following lemma.

Lemma 2.4.4.. [**Wel65, Kat74**] *The integral $\int_S f dm$, defined above, exists in each of the following cases:*

(a) $f \in C_{pc}(S, X)$;

(b) $f \in C_b(S, X)$ and $|m|_p$ is tight.

Proof. We need to show that, in each case, $\{S_\alpha : \alpha \in D\}$ is a Cauchy net in $\mathbb{R}$.

(a) Let $p \in cs(X)$ and $\varepsilon > 0$, and let $W = \{a \in X : p(a) \leq \varepsilon\}$. Since $f(S)$ is precompact, there exist $y_1, ..., y_n \in S$ such that

$$f(S) \subseteq \bigcup_{i=1}^{n} (f(y_i) + W).$$

Let $U_i = \{x \in S : f(x) - f(y_i) \in W\}$. Then each $U_i \in \mathcal{B}(S)$. Let

$$G_1 = U_1 \quad \text{and} \quad G_i = U_i \backslash \bigcup_{j=1}^{i-1} U_j (2 \leq i \leq n)$$

. By keeping those G_i's which are non-empty, we get, $\{G_1, ..., G_k\}$ say, a $\mathcal{B}(S)$-partition of S. Choose $x_i \in G_i$ and let $\alpha_0 = \{G_1, ..., G_k; \ x_1, ..., x_k\}$. Note that if x, y are in the same G_i,

$$p(f(x) - f(y)] = p(f(x) - f(y_i)] + p(f(y_i) - f(x)] \leq 2\varepsilon.$$

As before, let $S_{\alpha_0} = \sum_{i=1}^{k} m(G_i)(f(x_i))$. Then, for $\alpha_1, \alpha_2 \geq \alpha_0$, we have

$$|S_{\alpha_1} - S_{\alpha_2}| \leq |S_{\alpha_1} - S_{\alpha_0}| + |S_{\alpha_0} - S_{\alpha_2}|.$$

Suppose $\alpha_1 = \{F_1, ..., F_q; \ z_1, ..., z_q\}$, where each F_j is contained in some G_i and $z_j \in F_j$. Now

$$
\begin{aligned}
|S_{\alpha_1} - S_{\alpha_0}| &= \left| \sum_{j=1}^{q} m(F_j)(f(z_j)) - \sum_{i=1}^{k} m(G_i)(f(x_i)) \right| \\
&= \left| \sum_{j=1}^{q} m(F_j)(f(z_j)) - \sum_{i=1}^{k} \sum_{j, F_j \subseteq G_i} m(F_j)(f(x_i)) \right| \\
&= 2\varepsilon \left| \sum_{i=1}^{k} \sum_{j, F_j \subseteq G_i} m(F_j)[\varepsilon^{-1}(f(z_j) - f(x_i))] \right| \leq 2\varepsilon |m|_p (S).
\end{aligned}
$$

Similarly, we can prove that $|S_{\alpha_2} - S_{\alpha_0}| \leq 2\varepsilon |m|_p (S)$. Thus $|S_{\alpha_2} - S_{\alpha_1}| \leq 4\varepsilon |m|_p (S)$, and $\{S_\alpha : \alpha \in D\}$ is a Cauchy net in $\mathbb{R}$.

(b) Let $p \in cs(X)$ and $\varepsilon > 0$, and let $W = \{a \in X : p(a) \leq \varepsilon\}$. The collection $\mathcal{V} = \{f^{-1}(f(y) + W) : y \in S\}$ is a cover of S consisting of cozero sets. Labelling $\mathcal{V}$ as $\{V_\lambda : \lambda \in I\}$, we make I into a directed set by saying that $\lambda \geq \gamma \Leftrightarrow V_\lambda \subseteq V_\gamma$. Since $|m|_p$ is tight, there exist $y_1, ..., y_k \in S$ such that $|m|_p (S \backslash \bigcup_{i=1}^{k} V_{\lambda_i}) < \varepsilon$, where

$$V_{\lambda_i} = f^{-1}(f(y_i) + V) = \{x \in S : p(f(x) - f(y_i)) < 1\}.$$

Define $G_1 = V_{\lambda_1}, G_i = (V_{\lambda_i} \setminus \bigcup_{j=1}^{i-1} V_{\lambda_j}$ $(2 \leq i \leq k)$, and $G_{k+1} = S \setminus \bigcup_{i=1}^{k} V_{\lambda_i}$. Assuming that G_i's are non-empty, choose $x_i \in G_i$ and let $\alpha_0 = \{G_1, ..., G_{k+1}; x_1, ..., x_{k+1}\}$.

Let $\alpha_1, \alpha_2 \geq \alpha_0$. Suppose $\alpha_1 = \{F_1, ..., F_q; z_1, ..., z_q\}$ where each F_j is contained in some G_i and $z_j \in F_j$. Then

$$|S_{\alpha_1} - S_{\alpha_0}| \leq \left| \sum_{i=1}^{k} \sum_{j, F_j \subseteq G_i} m(F_j)(f(z_j) - f(x_i)) \right| +$$

$$+ \left| \sum_{j, F_j \subseteq G_{k+1}} m(F_j)(f(z_j)) \right| + \left| \sum_{j, F_j \subseteq G_{k+1}} m(F_j)(f(x_{k+1})) \right|$$

$$\leq \varepsilon(|m|_p (S) + 2\|f\|_p).$$

Similarly, we obtain $|S_{\alpha_2} - S_{\alpha_0}| \leq \varepsilon(|m|_p (S) + 2\|f\|_p)$. Consequently, $\{S_\alpha : \alpha \in D\}$ is a Cauchy net in $\mathbb{R}$. $\square$

Remark. Part (b) and (c) were proved in [**Fon74, Lemma 3.11**] assuming X a normed space.

Lemma 2.4..5. [**KR81**] *Let* $m \in M_p(S, X^*)$ $(p \in cs(X))$ *and* $f \in C_b(S, X)$. *If the integral* $\int_S f dm$ *exists, then*

$$\left| \int_S f dm \right| \leq \int_S (p \circ f) d|m|_p \leq \|f\|_p |m|_p (S).$$

Proof. For any $\varepsilon > 0$, There exist a $\mathcal{B}(S)$-partition, $\{F_i : 1 \leq i \leq n\}$ say, of S, and points $x_i \in F_i$ such that

$$\left| \int_S f dm \right| \leq \left| \sum_{i=1}^{n} m(F_i)(f(x_i)) \right| + \varepsilon$$

and

$$\left| \sum_{i=1}^{n} (p \circ f)(x_i) |m|_p (F_i) \right| \leq \int_S (p \circ f) d|m|_p + \varepsilon.$$

Let H_1 (resp. H_2) be the set of $i \in \{1, ..., n\}$ such that $p(f(x_i)) \neq 0$ (resp. $p(f(x_i)) = 0$). We note that, if $j \in H_2$, then $p(tf(x_i)) = 0$ for all $t > 0$. Then

$$\left| \int_S f dm \right| \leq \sum_{i \in H_1} (p \circ f)(x_i) \left| m(F_i) \left(\frac{f(x_i)}{(p \circ f)(x_i)} \right) \right|$$

$$+ \sum_{i \in H_2} \frac{\varepsilon}{|m|_p (S)} \left| m(F_i) \left(\frac{|m|_p (S) f(x_i)}{\varepsilon} \right) \right| + \varepsilon$$

$$\leq \sum_{i \in H_1} (p \circ f)(x_i) \left| m(F_i) \left(\frac{f(x_i)}{(p \circ f)(x_i)} \right) \right| + 2\varepsilon.$$

We note that, if $|m|_p (S) = 0$, then the inequality we are seeking to establish holds trivially. It follows that

$$\left| \int_S f dm \right| \leq \sum_{i \in H_1} (p \circ f)(x_i) |m|_p (F_i) + 2\varepsilon$$

$$\leq \int_S (p \circ f) d|m|_p + 2\varepsilon,$$

and so, since ε is arbitrary,

$$\left|\int_S f\,dm\right| \leq \int_S (p \circ f)\,d\,|m|_p.$$

The other inequality is straightforward to prove.

Note. It is easy to verify that, if $m \in M(\mathcal{B}(S), X^*)$, $g \in C_b(S)$, and $a \in X$, then

$$\int_S dm(g \otimes a) = \int_S g\,dm_a.$$

The Riesz-Kakutani type Theorems for the dual of $(C_{pc}(S, X), \beta)$

We first characterize the u-dual of $C_{pc}(S, X)$ via the integral representation, using the terminology of **Katsaras [Kat74]**.

Theorem 2.4.6.. [**Kat74**] $(C_{pc}(S, X), u)^* = M(\mathcal{B}(S), X^*)$ *via the linear isomorphism* $L \to m$ *given by*

$$L(f) = \int_S f\,dm \qquad (f \in C_{pc}(S, X)). \tag{\#2.17}$$

*Furthermore, if L is represented as in (**#2.17**) with $m \in M_p(\mathcal{B}(S), X^*)$ $(p \in cs(X))$, then $\|L\|_p = |m|_p(S)$, where*

$$\|L\|_p = \sup\{|L(f)| : f \in C_{pc}(S, X), \|f\|_p \leq 1\}.$$

Proof. Let $m \in M_p(\mathcal{B}(S), X^*)$ $(p \in cs(X))$, and suppose that L is the linear functional on $C_{pc}(S, X)$ defined by (**#2.17**). Then, by **Lemma 2.4.5**,

$$|L(f)| \leq \|f\|_p\,|m|_p(S) \leq |m|_p(S),$$

whenever $f \in C_{pc}(S, X)$ with $f(S) \subseteq W = \{a \in X : p(a) \leq 1\}$. Hence $L \in (C_{pc}(S, X), u)^*$. We now show that $\|L\|_p = |m|_p(S)$. Clearly, $\|L\|_p \leq |m|_p(S)$. For any $\varepsilon > 0$, there exist a $\mathcal{B}(S)$-partition $\{F_1, ..., F_n\}$ of S and a collection $\{a_1, ..., a_n\} \subseteq W$ such that

$$|m|_p(S) \leq \left|\sum_{i=1}^{n} m(F_i)(a_i)\right| + \varepsilon.$$

There exist zero sets $Z_i \subseteq F_i$ such that $|m_{a_i}|(Z_i \backslash F_i) < \frac{\varepsilon}{n}$. Since $C_b(S)$ separates the zero sets in S, the closures $\overline{Z}_1, ..., \overline{Z}_n$ in βS are pairwise disjoint. Since the cozero sets in βS form a base for open sets, there exist pairwise disjoint cozero sets $\hat{U}_1, ..., \hat{U}_2$ in βS with $\overline{Z}_i \subseteq \hat{U}_i$. Each $\hat{U}_i \cap S$ is a cozero set in S containing Z_i; there is a cozero set V_i in S with

$$Z_i \subseteq V_i \subseteq \hat{U}_i \cap S \text{ and } |m_{a_i}|(V_i \backslash Z_i) < \frac{\varepsilon}{n}.$$

Choose $h_i \in C_b(S)$ with $0 \leq h_i \leq 1$, $h_i(Z_i) = 1$, and $h_i = 0$ on $S \backslash V_i$. Let $h = \sum_{i=1}^{n} h_i \otimes a_i$. Then $h \in C_{pc}(S, X)$ and $\|h\|_p \leq 1$. Now

$$L(h) = \sum_{i=1}^{n} L(h_i \otimes a_i) = \sum_{i=1}^{n} \int_{Z_i} dm_{a_i} + \sum_{i=1}^{n} \int_{V_i \backslash Z_i} h_i\,dm_{a_i};$$

hence

$$\left| \sum_{i=1}^{n} m_{a_i}(F_i) \right| \leq \left| \sum_{i=1}^{n} m_{a_i}(F_i \backslash Z_i) \right| + \left| \sum_{i=1}^{n} m_{a_i}(Z_i) \right|$$

$$\leq \quad \varepsilon + |L(h)| + \sum_{i=1}^{n} |m_{a_i}| (V_i \backslash Z_i)$$

$$\leq \quad \|L\|_p + 2\varepsilon.$$

Thus $|m|_p(S) \leq \|L\|_p$.

Conversely, suppose that $L \in (C_{pc}(S, X), u)^*$. Then there exist a $p \in cs(X)$ and $r > 0$ such that

$$|L(f)| \leq r \|f\|_p \quad (f \in C_{pc}(S, X)). \tag{#2.18}$$

For each $a \neq 0$ in X, let $L_a(g) = L(g \otimes C_b(S))$. By (#**2.18**), we have

$$|L_a(g)| \leq r\|g\|p(a) \quad \text{for all } g \in C_b(S),$$

and so $L_a \in (C_b(S), u)^*$. Hence, by **Theorem 2.3.1**, there exists an $m_a \in M(\mathcal{B}(S))$ such that $L_a(g) = \int_S g \, dm_a$ $(g \in C_b(S))$ and $\|L_a\| = |m_a|(S)$.

For each $F \in \mathcal{B}(S)$, define $m(F)(a) = m_a(F)$ $(a \in X)$. Then $|m(F)(a)| \leq |m_a|(S) \leq rp(a)$, and consequently m is a finitely additive X^*-valued set function on $\mathcal{B}(S)$. Next $|m|_p(S) \leq r$, as follows. Let $\{F_1, ..., F_p\}$ be a $\mathcal{B}(S)$-partition of S and $\{a_1, ..., a_p\} \subseteq W = \{a \in X : p(a) \leq 1\}$, and let $\varepsilon > 0$. By using the argument as the one used earlier, there exists an $h \in C_{pc}(S, X)$ with $\|h\|_p \leq 1$ such that

$$\left| \sum_{i=1}^{p} m(F_i)(a_i) \right| \leq |L(h)| + 2\varepsilon \leq r + 2\varepsilon.$$

So $|m|_p(S) \leq r$ and hence $m \in M_p(\mathcal{B}(S), X^*)$.

Now for any $f = \sum_{i=1}^{k} f_i \otimes a_i$ $(f_i \in C_b(S), \ a_i \in X)$ in $C_b(S) \otimes X$,

$$L(F) = \sum_{i=1}^{k} L(f_i \otimes a_i) = \sum_{i=1}^{k} \int_S f_i \, dm_{a_i} = \sum_{i=1}^{k} \int_S dm(f_i \otimes a_i) = \int_S f \, dm.$$

Since $\int_x C_b(S) \otimes X$ is weakly dense in $(C_{pc}(S, X), u)$, the above holds for all $f \in C_{pc}(S, X)$.

Finally, m is unique, as follows. Suppose there is a $\mu \in M(\mathcal{B}(S), X^*)$ such that $L(f) = \int_S d\mu f$ for all $f \in C_{pc}(S, X)$. In particular, for any $g \in C_b(S)$ and $a \in X$, $\int_S dm(g \otimes a) = \int_S d\mu(g \otimes a)$. Hence $\int_S g \, dm_a = \int_S g \, d\mu_a$ for all $g \in C_b(S)$, and so by Theorem 2.3.1 , $m_a = \mu_a$. Thus, for any $F \in \mathcal{B}(S)$ and any $a \in X$, $m(F)(a) = m_a(F) = \mu_a(F)$. It follows that $m(F) = \mu(F)$, and so $m = \mu$. $\square$

The following representation theorem gives a characterization for the dual of $(C_b(S, X), \beta)$.

Theorem 2.4.7. [Kat75, KR81] $(C_b(S, X), \beta)^* = M_t(\mathcal{B}o(S), X^*)$ *via the linear isomorphism* $L \to m$ *given by*

$$L(f) = \int_S f dm \quad (f \in C_b(S, X)). \qquad (\#2.19)$$

Further, if L is represented as in (#2.19) with $m \in M_{t,p}(\mathcal{B}o(S), X^*)$ ($p \in cs(X)$), then

$$\|L\|_p = |m|_p(S).$$

Proof. Let $m \in M_{t,p}(\mathcal{B}o(S), X^*)$ ($p \in cs(X)$), and suppose that L is defined by (#2.19). Then, by **Lemma 2.4.2**, $|m|_p \in M_t(\mathcal{B}o(S))$. It follows from **Theorem 2.3.6** that the equation

$$L_p(g) = \int_S g d|m|_p \quad (g \in C_b(S)). \qquad (\#2.20)$$

defines a β-continuous linear functional L_p on $C_b(S)$. Hence there exists a $\varphi \in B_0(S)$, $0 \leq \varphi \leq 1$, such that $|L_p(g)| \leq 1$ whenever $g \in C_b(S)$ with $\|\varphi g\| \leq 1$. For any $f \in C_b(S, X)$ with $\|\varphi(p \circ f)\| \leq 1$, we have by (#2.20)) and **Lemma 2.4.5**,

$$|L(f)| \leq \int_S (p \circ \tilde{f}) d|m|_p = L_p(p \circ f) \leq 1.$$

Thus L is β-continuous.

Conversely, let $L \in (C_b(S, X), \beta)^*$. Then there exists a $\varphi \in B_0(S)$ and $p \in cs(X)$ such that

$$|L(f)| \leq 1 \text{ for all } f \in C_b(S, X) \text{ with } \|f\|_{p,\varphi} \leq 1. \qquad (\#2.21)$$

For each $a \neq 0$ in X, let $L_a(g) = L(g \otimes a)$ ($g \in C_b(S)$). It easily follows from (#2.21) that $L_a \in (C_b(S), \beta)^*$ and so by **Theorem 2.3.6**, there exists a unique $m_a \in M_t(S)$ such that

$$L_a(g) = \int_S g dm_a \quad (g \in C_b(S)).$$

For each $F \in \mathcal{B}o(S)$, $m(F)(a) = m_a(F)$ ($a \in X$). Since $\beta \leq u$, L is u-continuous and so there exists a $p \in cs(X)$ such that $|L(f)| \leq 1$ whenever $f \in C_b(S, X)$ with $\|f\|_p \leq 1$. Consider $g \in C_b(S)$ with $\|g\| \leq 1$. Then $p(g(x)a) \leq p(a)$ for all $x \in S$, and so $\|g \otimes a\|_p \leq 1$ Thus $\|L_a\| \leq p(a)$, and so from the inequalities

$$|m(F)(a)| \leq |m_a(F)| \leq \|m_a\| = \|L_a\| \leq p(a),$$

the continuity of $m(F)$ follows. Consequently m is a finitely additive X^*-valued set function on $\mathcal{B}o(S)$ such that, for each $a \neq 0$ in X, $m_a \in M_t(S)$.

Next $|m|_p(S) < \infty$, as follows. Let $\{F_j : j = 1, ..., n\}$ be a $\mathcal{B}o(S)$-partition of S and $a_1, ..., a_n \in W$, and let $\varepsilon > 0$. Each m_a is regular and so there exist compact sets $K_j \subseteq F_j$ with

$$|m_{a_j}|(F_j \backslash K_j) < \varepsilon/2n,$$

and open sets $V_j \supseteq K_j$ with

$$|m_{a_j}(F_j \backslash K_j)| < \varepsilon/2n \, for \, j = 1, ..., n;$$

since K_j's are disjoint compact sets and S is completely regular, the V_j's may be chosen pairwise disjoint. Choose functions $g_j (1 \leq j \leq n)$ in $C_b(S)$, $0 \leq g_j \leq 1$, such that $g_i = 1$ on K_j and $supp(g_j) \subseteq V_j$. Let $h = \sum_{i=1}^{n} g_j \otimes a_j$. Then $h \in C_{pc}(S, X)$ and $\|h\|_p \leq 1$, and so $|L(h)| \leq 1$. By using the above inequalities as in the proof of **Theorem 2.4.6**, we have $|m|_p (S) \leq \|L\|_p$. Thus $m \in M_t(\mathcal{B}o(S), X^*)$.

Consider any $f = \sum_{i=1}^{k} f_i \otimes a_i$ $(f_i \in C_b(S), \ a_i \in X)$ in $C_b(S) \otimes X$. Then

$$L(f) = \sum_{i=1}^{k} L(f_i \otimes a_i) = \sum_{i=1}^{k} f_i dm_{a_i} = \sum_{i=1}^{k} \int_S dm(f_i \otimes a_i) = \int_S f dm.$$

Since $C_b(S) \otimes X$ is dense in $(C_b(S, X), \beta)$, the above holds for all $f \in C_b(S, X)$. The proof for uniqueness of m and for $\|L\|_p = |m|_p (S)$ are the same as in Theorem 2.4.6 with slight modifications.

2.5   Notes and comments

The idea to face a linear functional against a measure via integration goes back probably to **F. Riesz [Rie09]**. This idea has stimulated a lot of work on duality theory. This chapter describes the structure of some duals for function spaces frequently used in integral representation theory. We beggin with the dual $C_0^*(S, X)$ of the the Banach space $C_0(S, X)$ (uniform norm) of all continuous functions $f : S \to X$, vanishing at infinity, where S is a locally compact space and X a Banach space. A theorem of **I. Singer [Sing57]**, settled for S compact, states that the topological dual $C_0^*(S, X)$ is isometrically isomorphic to the Banach space $r\sigma bv(S, X^*)$ of all regular vector measures of bounded variation on S, with values in the strong dual X^*. Actually the original proof of this theorem **[Sing57]** contains some gaps about the strong σ-additivity and regularity of the measure λ attached to the functional U. These gaps has been filled by **J. Gil de Lamadrid** in **[Lam66]**. Another proof using the Hahn-Banach theorem and measures on product spaces, can be found in **[Hen96]**. Here we have included a constructive detailed proof of Singer's theorem for S locally compact Hausdorff space, **[Mez09c]** which is, as far as we know, different from that supplied elsewhere **[Sing57; Hen96]**.

3

Integral Representations in Banach Spaces

The Integral Representation theorem of Riesz asserts that if $C(S)$ is the Banach space of all continuous functions $f : S \to \mathbb{R}$, on the compact space S, with the uniform norm, then for each linear continuous functional $T : C(S) \to \mathbb{R}$, there exists a unique bounded measure μ on the Borel σ-algebra $\mathcal{B}(S)$ of S such that:

$$Tf = \int_S f \, d\mu, f \in C(S),$$

where the representation is given by a Lebesgue integral with $\|T\| = |\mu|$, the variation of μ (see **Section 2.1**).

In this Chapter we will consider some extensions of this theorem for vector valued operators, with respect to appropriate integration processes.

All the set functions considered here are assumed, except otherwise stated, to be defined on the Borel σ-algebra $\mathcal{B}(S)$ of the compact set S.

The symbol X^* means the topological dual of the Banach space X and X^{**} its bidual. We will denote by $L(X, Y)$ the space of all linear bounded operators from X into the Banach space Y. Recall the definition of the Banach spaces $M_{r\sigma bv}(\mathcal{B}(S), X^*)$ and $M_{r\sigma bv}(\mathcal{B}(S))$given in **Definition 1.4.3, and Theorem 1.4.4.**

We know from Riesz theorem that $M_{r\sigma bv}(\mathcal{B}(S))$ is isometrically isomorphic to the dual $C^*(S)$ of $C(S)$.

3.1 Integral Representation of Bartle-Dunford-Schwartz

Theorem 3.1.1. *Let $T : C(S) \to X$. be an operator from $C(S)$ into a Banach space X. Then there exists a unique set function $\mu : \mathcal{B}(S) \to X^{**}$ such that:*

(a) $\mu(\bullet) \, x^* \in M_{r\sigma bv}(\mathcal{B}(S))$, *for each $x^* \in X^*$.*

(b) *the mapping $x^* \to \mu(\bullet) \, x^*$ from X^* into $M_{r\sigma bv}(\mathcal{B}(S))$ is weak* continuous with respect to the topologies $\sigma(X^*, X)$ on X^* and $\sigma(C^*(S), C(S))$ on $C^*(S)$.*

(c) $x^* Tf = \int_S f(s) \, d\mu(s) \, x^*, f \in C(S), x^* \in X^*.$

(d) $\|T\| = \|\mu\|(S)$, *the semi-variation of μ at S (**Definition 1.1.3**).*

Conversely, if μ is a set function from $\mathcal{B}(S)$ into X^{**} satisfying **(a)** *and* **(b)**, *then equation* **(c)** *defines an operator* $T : C(S) \to X$ *with norm given by* **(d)**, *and such that* $T^* x^* = \mu(\bullet)\, x^*$.

Proof: In what follows we make use of the identication between the dual $C^*(S)$ of $C(S)$ and the Banach space $M_{r\sigma bv}(\mathcal{B}(S))$ of all regular real measures with bounded variation. Fix E in $\mathcal{B}(S)$. and define $\varphi_E : C^*(S) \to \mathbb{R}$, by $\varphi_E(\lambda) = \lambda(E)$, $\lambda \in C^*(S)$. It is clear that $|\varphi_E(\lambda)| \leq |\lambda|$ (variation norm) so that $\varphi_E \in C^{**}(S)$ and $\|\varphi_E\| \leq 1$. Next, define $\mu : \mathcal{B}(S) \to X^{**}$ by

$$\mu(E) = T^{**}(\varphi_E),\, E \in \mathcal{B}(S).$$

It is easy to check that μ is additive. Moreover if $T^* : X^* \to C^*(S)$ is the adjoint of T, then for $x^* \in X^*$, $T^* x^* = \lambda_{x^*} \in C^*(S)$ and

$$\lambda_{x^*}(E) = \varphi_E(\lambda_{x^*}) = \varphi_E(T^*(x^*)) = T^{**}\varphi_E x^*.$$

From the definition of μ, we deduce that $\lambda_{x^*}(E) = \mu(E)\, x^*$ and $T^* x^* = \mu(\bullet)\, x^*$. This proves **(a)** and **(b)**, where **(b)** comes from the fact that the adjoint T^* is weak* continuous. By Riesz theorem

$$T^* x^*(f) = \int_S f\, d\mu(\bullet)\, x^*$$

and, since $T^* x^*(f) = x^* T(f)$, we get **(c)**. Finally a straightforward computation gives **(d)**.

Conversely, let μ be a set function from $\mathcal{B}(S)$ into X^{**} satisfying **(a)** and **(b)**. Then for each $f \in C(S)$, the mapping $x^* \to \int_S f\, d\mu(\bullet)\, x^*$ of X^* into $\mathbb{R}$ is $\sigma(X^*, X)$-continuous; consequently there is a unique vector x_f in X such that

$$\int_S f\, d\mu(\bullet)\, x^* = x^*(x_f).$$

Define $T : C(S) \to X$ by $Tf = x_f$, $f \in C(S)$. It is not difficult to check that T satisfies **(c)** and **(d)**. $\square$

Now it is natural to ask when the set function μ takes its values in $\gamma(X)$, where $\gamma : X \to X^{**}$ is the canonical embedding. As we will see presently, this will be true if the operator T is weakly compact.

Definition 3.1.2. Let X, Y be Banach spaces. A linear bounded operator $T : X \to Y$ is said to be **weakly compact** if for any bounded set B in X, the weak closure of TB is compact in the weak topology of Y.

Let us note the following facts about weakly compact operators [**DS58,Chap.VI**]

Theorem 3.1.3. *A linear bounded operator* $T : X \to Y$ *is weakly compact if and only if* $T^{**} X^{**}$ *is contained in the natural embedding* $\gamma(Y)$ *of* Y *into* Y^{**}.

Theorem 3.1.4. *An operator* T *is weakly compact if and only if its adjoint* T^* *is weakly compact.*

Proof: See [**DS58, Chap.VI**].

Now we are in a position to state the integral representation of a weakly compact operator.

Theorem 3.1.5. *Let $T : C(S) \to X$. be a weakly compact operator from $C(S)$ into a Banach space X. Then there exists a unique vector measure $\lambda : \mathcal{B}(S) \to X$ such that:*

$(a^{'})$ $x^{*}\lambda \in M_{r\sigma bv}(\mathcal{B}(S))$, for each $x^{*} \in X^{*}$.

$(b^{'})$ $Tf = \int_{S} f\, d\lambda$, for all $f \in C(S)$

$(c^{'})$ $T^{*}x^{*} = x^{*}\lambda$, for each $x^{*} \in X^{*}$.

$(d^{'})$ $\|T\| = \|\lambda\|(S)$, the semi-variation of λ at S.

Conversely if λ is a vector measure on $\mathcal{B}(S)$ with values in X, satisfying $(a^{'})^{'}$, then the operator T defined by $(b^{'})$ is a weakly compact operator from $C(S)$ into X whose adjoint is given by $(c^{'})$ and whose norm is given by (d).

Proof: Let $\mu : \mathcal{B}(S) \to X^{**}$ be the set function introduced in **Theorem 3.1.1** by $\mu(E) = T^{**}(\varphi_{E})$, $E \in \mathcal{B}(S)$. Since T is weakly compact, $T^{**}X^{**}$ is contained in the natural embedding $\gamma(X)$ of X into X^{**} by **Theorem 3.1.3.** Therefore $\mu(E) \in \gamma(X)$ for all $E \in \mathcal{B}(S)$ and this allows to define $\lambda : \mathcal{B}(S) \to X$ by

$$\lambda(E) = \gamma^{-1}\mu(E).$$

Then $\lambda(E) \in X$ and for $x^{*} \in X^{*}$:

$$x^{*}\lambda(E) = x^{*}\gamma^{-1}\mu(E) = \gamma\left(\gamma^{-1}\mu(E)\right)x^{*} = \mu(E)\,x^{*},$$

where the second equality comes from the definition of γ. This shows that $x^{*}\lambda \in M_{r\sigma bv}(\mathcal{B}(S))$, for each $x^{*} \in X^{*}$. By **Theorem 1.1.5,** λ is a vector measure on $\mathcal{B}(S)$ into X. From **Theorem 3.1.1 (c)** we get

$$x^{*}Tf = \int_{S} f(s)\, d\mu(s)\, x^{*} = \int_{S} f(s)\, dx^{*}\lambda(s),\ f \in C(S), x^{*} \in X^{*}.$$

Now apply **Theorem 3.1.1(c)** to get $x^{*}Tf = x^{*}\int_{S} f(s)\, d\lambda(s)$, for each $x^{*} \in X^{*}$. This yields $Tf = \int_{S} f\, d\lambda$, by the Hahn-Banach Theorem, so $(b^{'})$ is proved. On the other hand we have

$$x^{*}Tf = \int_{S} f(s)\, dx^{*}\lambda(s) = T^{*}x^{*}(f),$$

hence $T^{*}x^{*} = x^{*}\lambda$, by Riesz theorem, whence $(c^{'})$. Finally $(d^{'})$ is immediate from part (d) of **Theorem 3.1.1.**

Conversely if λ is a vector measure on $\mathcal{B}(S)$ with values in X, satisfying $(a^{'})$ and if T is defined by $(b^{'})$, then T is linear bounded (**Theorem 1.2.3** (a)) and for each $x^{*} \in X^{*}$ we have $T^{*}x^{*} = x^{*}\lambda$. By **[BDS55, Lemma 2.3]** T^{*} is weakly compact and so, **Theorem 3.1.4,** T is weakly compact. $\square$

Note that for a locally compact space S, **Theorem 3.1.5** has been extended by **Kluvanek** in **[Klu67, Lemma 2]**.

In the next representation theorem, we need the structure of the topological dual $C_{0}^{*}(S, X)$ of the Banach space $C_{0}(S, X)$. This is given in **Theorem**

2.2.1 of **Chapter 2:** there is an isometric isomorphism between the Banach spaces $C_0^*(S, X)$ and $M_{r\sigma bv}(\mathcal{B}(S), X^*)$, where to the functional U in $C_0^*(S, X)$ corresponds the measure λ in $M_{r\sigma bv}(\mathcal{B}(S), X^*)$ via the formula:

$$\begin{aligned} U(f) &= \int_S f \, d\lambda, \, f \in C_0(S, X); \\ \|U\| &= |\lambda| \end{aligned}$$

Now we turn to the one of the most general representation theorem.

3.2   Representation Theorem of Dinculeanu-Singer

In this section S is a compact space equipped with its Borel $\sigma-$algebra $\mathcal{B}(S)$. It is clear that in this case we have $C_0(S, X) = C(S, X)$.

The integration process needed in this section has been described in **Section 1.5**, the notations are those used there. It is performed for vector valued functions $f : S \to X$, with respect to an additive set function on $\mathcal{B}(S)$ with values in $L(X, Y^{**})$ and gives the integral $\int_S f \, dG$ as a vector of Y^{**}. Recall that to each additive $G : \mathcal{B}(S) \to L(X, Y^{**})$ there is associated a family $\{G_{y^*}, \, y^* \in Y^*\}$ of additive X^*-valued set functions given by:

$$G_{y^*}(A).x = G(A)x(y^*), \quad A \in \mathcal{B}(S), \, x \in X.$$

With this in mind, we have:

Theorem 3.2.1. *Every linear bounded operator* $T : C(S, X) \to Y$ *determines a unique set function* $G : \mathcal{B}(S) \to L(X, Y^{**})$ *such that:*

(i) G is finetely additive and with finite semivariation $\widetilde{G}$.

(ii) The set function G_{y^*} is a vector measure in $M_{r\sigma bv}(\mathcal{B}(S), X^*)$ for each $y^* \in Y^*$.

(iii) The function $y^* \to G_{y^*}$ is weak* continuous with the $\sigma(Y^*, Y)$-topology on Y^* and the $\sigma(C^*(S, X), C(S, X))-$topology on $C^*(S, X)$.

(iv) $Tf = \int_S f \, dG, \, f \in C(S, X)$, which really means that $\gamma Tf = \int_S f \, dG$, where $\gamma : Y \to Y^{**}$ is the canonical embedding.

*(v) $\|T\| = \widetilde{G}(S)$ (semivariation of G defined in **Section 1.2**).*

(vi) $T^* y^* = G_{y^*}$, for all $y^* \in Y^*$.

Proof: Consider the adjoint operator $T^* : Y^* \to C^*(S, X)$. From **Theorem 3.2.1,** for each $y^* \in Y^*$, $T^* y^*$ is a vector measure on $\mathcal{B}(S)$ with values in X^*, which we will denote by μ_{y^*}. Thus we have for $f \in C(S, X)$,

$$T^* y^*(f) = y^* Tf = \int_S f \, d\mu_{y^*}, \text{ and } \|T^* y^*\| = |\mu_{y^*}|.$$

Next fix $x \in X$, $A \in \mathcal{B}(S)$ and define $G(A)x : Y^* \to \mathbb{R}$ by $G(A)x(y^*) = \mu_{y^*}(A)(x)$. Then it is easy to check that $G(A)x \in Y^{**}$ and that we have $\|G(A)x\| \leq \|T\| . \|x\|$. This allows us to define, for A fixed in $\mathcal{B}(S)$, $G(A):$

$X \to Y^{**}$ by $x \to G(A)x$. Then $G(A)$ is linear and bounded with $\|G(A)\| \leq \|T\|$, for all $A \in \mathcal{B}(S)$. Furthermore, the set function $A \to G(A)$ from $\mathcal{B}(S)$ into $L(X, Y^{**})$ is additive and satisfies:

$$G(A)x(y^*) = G_{y^*}(A)(x) = \mu_{y^*}(A)(x), A \in \mathcal{B}(S), y^* \in Y^*, x \in X.$$

Moreover we have, from **Proposition 1.4.6**,

$$\widetilde{G}(S) = \sup\{|G_{y^*}|, \ \|y^*\| \leq 1\}.$$

So we deduce that $\widetilde{G}(S) = \sup\{|\mu_{y^*}|, \ \|y^*\| \leq 1\}$. Since $\|T^*y^*\| = |\mu_{y^*}|$, we get

$$\widetilde{G}(S) = \sup\{\|T^*y^*\|, \ \|y^*\| \leq 1 = \|T^*\| = \|T\|\}.$$

Let us observe that:

(i) is satisfied by the definition of G.

(ii) is a consequence of the relation $G_{y^*} = \mu_{y^*}$.

(vi) is true since $T^*y^* = \mu_{y^*} = G_{y^*}$.

(iii) is valid by (vi) and the weak*continuity of the adjoint.

(v) is proved by the computation above involving $\widetilde{G}(S)$.

(vi) Here we use the integration process described in **Section 1.5**. Consider the space $F_m(S, X)$ of measurable functions $f : S \to X$. By **Proposition 1.5.5**, we have $C(S, X) \subseteq F_m(S, X)$, so $\int_S f \, dG$ is well defined for $f \in C(S, X)$, and is in Y^{**}, since G takes its values in $L(X, Y^{**})$. Put for a moment $Uf = \int_S f \, dG$ and observe that for $y^* \in Y^*$,

$$Uf(y^*) = \int_S f \, dG_{y^*}$$

(check the formula for f simple and extend to all $f \in F_m(S, X)$, using standard methods.). But we have also

$$T^*y^*(f) = y^*Tf = \int_S f \, dG_{y^*},$$

hence $y^*Tf = Uf(y^*)$, for all $f \in C(S, X)$ and $y^* \in Y^*$. Since $Tf \in Y$, we have $y^*Tf = \gamma Tf(y^*)$, where $\gamma : Y \to Y^{**}$ is the canonical embedding. Consequently $\gamma Tf(y^*) = Uf(y^*)$, and then $\gamma Tf = Uf$ for all $f \in C(S, X)$, proving (iv). $\square$

3.2.2 Remark: The set function G above is not σ−additive in general. To get σ−additivity, needs additional assumption on the operator T. It has been proved by **Dobrakov** in **[Dob71]**, that the representing measure G of the operator $T : C(S, X) \to Y$ has all its values in $L(X, Y)$ if and only if for each $x \in X$ the operator $T_x : C(S) \to Y$ defined by

$$T_x f = T(x.f), f \in C(S),$$

is weakly compact. In this case, G is σ−additive in the strong operator topology of $L(X, Y)$. In the next section we will meet a class of operators having such property.

3.3 Bounded Operators and Bochner Integral

In this section, we introduce a class of bounded operators $T : C(S, X) \to X$, which admit a representation by Bochner integral with respect to a scalar measure on $\mathcal{B}(S)$. For the construction of the Bochner integral and its properties, we refer the reader to [**HP57**], see also **Section 1.3**.

Definition 3.3.1 Let μ be a scalar measure with bounded variation on $\mathcal{B}(S)$ and let us consider **the operator** $I_\mu : C(S, X) \to X$ introduced in **Theorem 1.3.11**,

$$f \in C(S, X), I_\mu f = \int_S f \, d\mu$$

where the integral is in the sense of Bochner. Note that, since the function $s \to \|f(s)\|$ is bounded and the measure μ has finite variation, the operator I_μ is well defined, by the integrability criterion of Bochner integral.

For each $x^* \in X^*$ let $U_{x^*} : C(S, X) \to C(S)$ be the linear bounded operator given by

$$U_{x^*} f = x^* \circ f,$$

where $(x^* \circ f)(s) = x^*(f(s))$, $f \in C(S, X)$, $s \in S$.

We collect some facts about U_{x^*} for later use:

Proposition 3.3.2. *(1) For each $x^* \in X^*$, we have $\|U_{x^*}\| = \|x^*\|$. Moreover there exists $z^* \in X^*$ such that for every $h \in C(S)$, there is a solution $f \in C(S, X)$ of the equation $U_{z^*} f = h$ with $\|f\| = \|h\|$.*

(2) If $V : C(S) \to \mathbb{R}$ is linear and bounded then we have:

$$\|V\| = \sup\{\|V \circ U_{x^*}\|, \ \|x^*\| \le 1\}.$$

Proof: The proposition is a consolidated form of [**Mez02, Lemmas 2.2, 2.3**]. $\square$

All what we need about I_μ is the following (see also **Theorem 1.3.10**):

Proposition 3.3.3. *(1) I_μ is a linear bounded operator from $C(S, X)$ into X. Moreover, if $U : X \to E$, is a bounded operator from X into the Banach space E, then we have*

$$U(I_\mu f) = I_\mu(U f),$$

for all $f \in C(S, X)$, where Uf is the function in $C(S, E)$ given by $(Uf)(s) = U(f(s))$, for $s \in S$.

(2) $\|I_\mu\| = |\mu|$ (the variation of μ).

Proof: (1) This follows from **Theorem 3.1.1**.

(2) This is a part of **Theorem 1.3.10**. $\square$

Now let us consider a bounded operator $T : C(S, X) \to X$, and ask the question of the existence of a scalar measure μ on $\mathcal{B}(S)$ such that $Tf = I_\mu f$, for all $f \in C(S, X)$. In what follows we introduce a class of operators $T : C(S, X) \to X$ for which this problem does have a positive answer.

Definition 3.3.4. Let C_{XX} be the class of linear bounded operators $T : C(S, X) \to X$ which satisfy the following condition $(\#C)$:

$$x^* \circ f = y^* \circ g \Longrightarrow x^*Tf = y^*Tg \quad x^*, y^* \in X^*, \quad f, g \in C(S, X). \quad (\#C)$$

In some sense, the operators of the class C_{XX} preserve the continuous functionals of X. It is easy to check that C_{XX} is a closed subspace of the Banach space $L\left(C(S, X), X\right)$, endowed with the uniform norm. Also let us observe that for each scalar measure μ with bounded variation on $\mathcal{B}(S)$, the operator I_μ is in C_{XX} (take $E = \mathbb{R}$ and $U = U_{x^*}$ in **Proposition 3.3.3 (1)**).

The outstanding fact about the class C_{XX} is contained in the following theorem which will be essential for the integral representation.

Theorem 3.3.5. [Mez02] *There is an isometric isomorphism between the Banach space C_{XX} and the topological dual $C^*(S)$ of $C(S)$, for each non trivial Banach space X. In other words, there exists a linear bijective mapping $\varphi : C_{XX} \to C^*(S)$ such that*

$$\|\varphi(T)\| = \|T\|, \text{ for all } T \in C_{XX}.$$

Proof: We show how to construct the mapping $\varphi : C_{XX} \to C^*(S)$, with the given properties. Let $T \in C_{XX}$ and $h \in C(S)$. By **3.3.2 (1)** there is $z^* \in X^*$ and $f \in C(S, X)$ such that

$$U_{z^*} f = h \text{ and } \|f\| = \|h\|.$$

Next define $V : C(S) \to \mathbb{R}$, $V(h) = z^*Tf$. Then V is well defined because of condition $(\#C)$ imposed to the operator T; moreover V is linear and bounded, i.e $V \in C^*(S)$. Let us define $\varphi : C_{XX} \to C^*(S)$, by

$$\varphi(T) = V, T \in C_{XX}$$

It is clear from this construction that φ is linear. Furthermore we have:

$$V \circ U_{x^*} = x^* \circ T, \qquad \forall x^* \in X^*, \qquad\qquad (\#3.1)$$

Indeed, for $f \in C(S, X)$ and $x^* \in X^*$, $U_{x^*} f$ is in $C(S)$, so by **Proposition 3.3.2(1)** there exists $g \in C(S, X)$ such that

$$U_{x^*} f = x^* \circ f = z^* \circ g.$$

Therefore:

$$\begin{aligned} V \circ U_{x^*}(f) &= V(x^* \circ f) = z^* \circ Tg \text{ (by the definition of } V) \\ &= x^* \circ Tf \text{ (by condition } \#C) \end{aligned}$$

since f is arbitrary, ($\#\mathbf{3.1}$) is proved.

On the other hand, we have from **Proposition 3.3.2**(2),

$$\|V\| = \sup\{\|V \circ U_{x^*}\|, \ \|x^*\| \le 1\},$$

and using ($\#\mathbf{3.1}$) we get

$$\|V\| = \sup\{\|x^* \circ T\|, \ \|x^*\| \le 1\} = \|T^*\| = \|T\|.$$

This proves that $\|V\| = \|\varphi(T)\| = \|T\|$, that is φ is an isometry.

To achieve the proof, we construct $\theta : C^*(S) \to C_{XX}$, which will be the inverse of φ. If $V \in C^*(S)$, then, by **Riesz theorem**, there exists a bounded real measure μ on $\mathcal{B}(S)$ such that

$$Vh = \int_S h\, d\mu \quad \text{and} \quad \|V\| = |\mu| \text{ for all } h \in C(S).$$

Then, define θ by $\theta(V) = I_\mu$. We know that $I_\mu \in C_{XX}$, so θ is well defined and is an isometry from $C^*(S)$ into C_{XX}, since $\|\theta(V)\| = \|I_\mu\| = |\mu| = \|V\|$. We prove that θ is the inverse of φ. Let $T \in C_{XX}$, with $\varphi(T) = V \in C^*(S)$. Let μ be the measure associated to V as before. By definition we have

$$\theta(V) f = I_\mu(f) = \int_S f\, d\mu$$

and for each $x^* \in X^*$,

$$
\begin{aligned}
x^* \theta(V) f &= x^* \int_S f\, d\mu = \int_S x^* \circ f\, d\mu \\
&= V(x^* \circ f) = x^* T f \qquad \text{(from ($\#$3.1), $V = \varphi(T)$)}
\end{aligned}
$$

since x^* is arbitrary, we deduce from **Hahn-Banach theorem** that $\theta(V) f = T f$ and consequently $\theta \circ \varphi(T) = T$. Similarly we have $\varphi \circ \theta(V) = V$, all $V \in C^*(S)$. $\square$

As a consequence of this theorem, we give a representation of an operator in the class C_{XX} by mean of a **Bochner integral**.

Theorem 3.3.6. *Let $T : C(S, X) \to X$ be an operator in the class C_{XX}. Then there is a unique bounded real measure μ on $\mathcal{B}(S)$ such that*
(i) $T(f) = \int_S f\, d\mu$, *for all $f \in C(S, X)$.*
(ii) $\|T\| = |\mu|$.

Proof: We use the transformations φ and θ introduced in the proof of **Theorem 3.3.5**. Put

$$\varphi(T) = V \in C^*(S) \text{ and } Vh = \int_S h\, d\mu \text{ for } h \in C(S).$$

Appealing to the relation $(*)$ used in the proof of **Theorem 3.3.5**, we get:

$$
\begin{aligned}
V(x^* \circ f) &= x^* \circ T f = \int_S x^* \circ f\, d\mu \\
&= x^* \int_S f\, d\mu, \ f \in C(S, X) \text{ and } x^* \in X^*.
\end{aligned}
$$

Consequently

$$x^* \circ T f = x^* \int_S f\, d\mu \text{ for all } x^* \in X^*,$$

which gives $T f = \int_S f\, d\mu$. $\square$

3.4   Bochner Integral Representation

If T is an operator in the class C_{XX}, then T admits at least two integral representations. These are:

$$Tf = \int_S f\, dG, \ f \in C(S, X), \qquad (\#3.2)$$

where $G : \mathcal{B}(S) \to L(X, X^{**})$ is an additive set function (Theorem **3.2.1**).

$$Tf = \int_S f\, d\mu, \ f \in C(S, X), \qquad (\#3.3)$$

where $\mu : \mathcal{B}(S) \to \mathbb{R}$ is a scalar measure of bounded variation (**Theorem 3.3.6**).

It is known, [**DU77, Din67**], that some properties relevant to the nature of the operator T are reflected by the structure of the set function G given in **Theorem 3.2.1**. Thus it is important to investigate the structure of G, and we will do this by comparing formulas $(\#3.2)$, $(\#3.3)$, for an operator T in the class C_{XX}.

For each $y^* \in X^*$, recall the set function $G_{y^*} : \mathcal{B}(S) \to X^*$, is given by:

$$G_{y^*}(A)x = G(A)x\,(y^*), \ A \in \mathcal{B}(S), x \in X.$$

Then we know from **Theorem 3.2.1**, that G_{y^*} is regular and $\sigma-$additive. Next define the family of measures $\Lambda^x_{y^*} : \mathcal{B}(S) \to X^*$, by:

$$\Lambda^x_{y^*}(A) = G_{y^*}(A)x, \ A \in \mathcal{B}(S), x \in X, y^* \in X^*.$$

It is clear that $\Lambda^x_{y^*}$ are regular. Moreover we have:

Theorem 3.4.1. [**Mez02**] *Let* $G : \mathcal{B}(S) \to L(X, X^{**})$ *be satisfying Theorem **3.2.1**, and let* $\mu : \mathcal{B}(S) \to \mathbb{R}$ *be as in* $(\#3.3)$*. Then we have:*

$$G(\bullet) = \mu(\bullet)\,.\gamma \qquad (\#3.4)$$

where the relation has the meaning that $G(A).x = \mu(A).\gamma(x)$*, for all* $A \in \mathcal{B}(S)$ *and* $x \in X$*;* $\gamma : X \to X^{**}$*, is the canonical embedding, with*

$$\gamma(x)\,y^* = y^*(x), x \in X, y^* \in X^*.$$

Note that $(\#3.4)$ is equivalent to:

$$G_{y^*}(A)x = \mu(A).y^*(x). \qquad (\#3.5)$$

Proof: Let $f \in C(S, X)$ having the form $f(s) = g(s)\,.x$, with $g : S \to \mathbb{R}$, continuous and where x is fixed in X. Then from $(\#3.2)$ and $(\#3.3)$ we get:

$$Tf = \int_S g(s)\,.x\, dG = \int_S g(s)\,.x\, d\mu.$$

Applying $y^* \in X^*$ to Tf, yields:

$$y^*Tf = T^*y^*(f) = \int_S g(s)\,.x\, dG_{y^*},$$

where the second equality comes from **Theorem 3.2.1(vi)**. But $\int_S g\,(s)\,.x\,d\,G_{y^*} = \int_S g\,(s)\,.d\,\Lambda^x_{y^*}$, as may be seen, first by taking g a simple function and extending with standard integration tools. Going back to the Bochner integral form of Tf and applying $y^* \in X^*$, we obtain

$$y^*Tf = y^* \int_S g\,(s)\,.x\,d\,\mu = \int_S g\,(s)\,.y^*\,(x)\,d\,\mu \;= \int_S g\,(s)\,d\,y^*\,(x)\,.\mu.$$

Consequently, from the other form of y^*Tf , we get

$$\int_S g\,(s)\,d\,y^*\,(x)\,.\mu = \int_S g\,(s)\,.d\,\Lambda^x_{y^*},$$

for all continuous $g : S \to \mathbb{R}$. Since both measures $y^*\,(x)\,.\mu$ and $\Lambda^x_{y^*}$ are regular, we deduce that $y^*\,(x)\,.\mu = \Lambda^x_{y^*}$, by **Riesz theorem**. But $\Lambda^x_{y^*}\,(A) = G_{y^*}(A)x$, so this yields (**#3.5**) and then (**#3.4**) is proved. $\square$

Corollary: *The representing measure G, given by Dinculeanu-Singer Theorem for an operator T in the class C_{XX}, is $\sigma-$additive in the uniform topology of $L\,(X, X^{**})$.*

3.5   Notes and comments

In this chapter we considered three major representation theorems. The first one is the famous theorem of Bartle-Dunford Schwartz [**BDS55**], representing general $X-$valued operators on $C(S)$. In this representation, the class of weakly compact operators brings the important improvement of being represented by integrals against vector measures with values in the Banach space X.

The second theorem, due to **Dinculeanu-Singer** [**Din67**], for bounded operators $T : C(S, X) \to Y$, is one of the most general representation theorems. It goes through the structure of the topological dual $C^*\,(S, X)$ of the function space $C(S, X)$, given in [**Sing57**], and generalized in [**Mez09c**], see **Chapter 2** for details The aim of the third representation theorem is the construction of a class of operators from $C(S, X)$ into X, characterized by their Bochner form, given in [**Mez02**]. We bring the attention of the reader to the intersting comparison made in **Theorem 3.4.1**, between Dinculeanu-Singer and Bochner integral representations whenever such representation coexit for the same operator.

4

Integral Representations in LCSs

Now, the objective is to go beycnd the Banach space setting, to a TVS context. The aim of this Chapter is to get integral representation theorems of bounded operators by weak integrals, in an appropriate framework of TVS.

In all this Chapter, unless otherwise stated, S will be a topological space and μ a real measure of bounded variation on the Borel $\sigma-$algebra $\mathcal{B}(S)$. Also X will be a locally convex Hausdorff space with topological dual X^*, and for $\theta \in X^*, x \in X$, we denote by $\langle \theta, x \rangle$ the functional duality between X and X^*.

4.1 Integral Representation by Pettis Integral

Suppose that S is a locally compact space and let X be a locally convex TVS. We denote by $C_0(S, X)$ the set of all continuous functions $f : S \to X$, vanishing outside a compact set of S, put $C_0(S, X) = C_0(S)$ if $X = \mathbb{R}$. We are interested in representing linear bounded operators $T : C_0(S, X) \to X$, by means of weak integrals against scalar measures on the Borel $\sigma-$algebra $\mathcal{B}(S)$ of S. Before handling more closely this problem, we need some preliminary facts about the space $C_0(S, X)$.

Definition 4.1.1. If K is a compact set in S, let $C(S, K, X)$ be the set of all continuous functions $f : S \to X$, vanishing outside K. It is clear that $C(S, K, X)$ is a linear subspace of $C_0(S, X)$. We equip $C(S, K, X)$ with the topology τ_K generated by the family of seminorms $\{\tilde{p}_{\alpha,K}\}$:

$$\tilde{p}_{\alpha,K}(f) = \sup_{t \in K} p_\alpha(f(t)), \quad f \in C(S, K, X), \qquad (\#4.1)$$

where $\{p_\alpha\}$ is the family of seminorms generating the locally convex topology of X. The topology τ_K is the topology of uniform convergence on K.

Next let us observe that $C_0(S, X) = \bigcup_K C(S, K, X)$, the union being performed over all the compact subsets K of S. On the other hand if $K_1 \subseteq K_2$, then the natural embedding $i_{K_1 K_2} : C(S, K_1, X) \to C(S, K_2, X)$ is continuous. This allows one to provide the space $C_0(S, X)$ with the inductive topology τ induced by the subspaces $(C(S, K, X), \tau_K)$. The facts we need about the space $(C_0(S, X), \tau)$ are well known (see **Appendix A 1** for details):

Proposition 4.1.2. (a) *The space $(C_0(S,X),\tau)$ is locally convex Hausdorff and for each compact K, the relative topology of τ on $C(S,K,X)$ is τ_K, i.e the canonical embedding $i_K : (C(S,K,X),\tau_K) \to (C_0(S,X),\tau)$ is continuous.*

(b) *Let $T : C_0(S,X) \to V$ be a linear operator of $C_0(S,X)$ into the locally convex Hausdorff space V, then T is continuous if and only if the restriction $T \circ i_K$ of T to the subspace $C(S,K,X)$ is continuous for each compact K.*

Definition 4.1.3. For each θ in the topological dual X^* of X and for each function $f \in C_0(S,X)$, define the function $U_\theta f$ on S by $U_\theta f(s) = \theta(f(s)) = \langle \theta, f(s) \rangle$. Then U_θ sends $C_0(S,X)$ into $C_0(S)$. Recall that $C_0(S)$ is equipped with the uniform norm.

Lemma 4.1.4. *The operator U_θ is linear and bounded. Moreover for each $\theta \neq 0$, U_θ is onto.*

Proof: First it is clear that $U_\theta f \in C_0(S)$. Now by **Proposition 4.1.2(b)**, we have to show that for each compact set $K \subseteq S$ the operator $U_\theta \circ i_K : C(S,K,X) \to C_0(S)$ is bounded. Since θ is bounded, there is a seminorm p_α on X and a constant M such that $|\theta(x)| \leq M\, p_\alpha(x)$ for all $x \in X$. So we have $|\theta(f(s))| \leq M\, p_\alpha(f(s))$ if $f \in C(S,K,X)$, and $U_\theta \circ i_K(f)(s) = \theta(f(s))$, $s \in S$; it follows that

$$\|U_\theta f\| = \sup_{s \in K} |\theta(f(s))| \leq M. \sup_{s \in K} p_\alpha(f(s)). \qquad (\#4.2)$$

Since by $(\#4.1)$, the right side of this inequality is $M\, \widetilde{p}_{\alpha,K}(f)$, we deduce that U_θ is continuous. Now suppose $\theta \neq 0$. Then there exists $x \in X$ such that $x \neq 0$ and $\theta(x) \neq 0$. It is clear that we can assume $\theta(x) = 1$. Now let $h \in C_0(S)$ and define $f : S \to X$ by $f(t) = h(t).x$, then $f \in C_0(S,X)$ and we have $U_\theta(f)(s) = U_\theta(h(s)x) = h(s)$, because $\theta(x) = 1$. It follows that U_θ is onto. $\square$

In what follows we deal with the representation of bounded operators $T : C_0(S,X) \to X$, by weak integrals in the sense of the definition:

Definition 4.1.5. We say that a bounded operator $T : C_0(S,X) \to X$, has a **Pettis integral form** if there exist a scalar measure of bounded variation μ on $\mathcal{B}(S)$ such that, for every continuous functional θ in X^*, we have:

$$\langle \theta, Tf \rangle = \int_S \langle \theta, f \rangle \, d\mu, \quad f \in C_0(S,X). \qquad (\#P)$$

Definition 4.1.6. Let us denote by $\mathcal{P}$ the class of all bounded operators $T : C_0(S,X) \to X$ satisfying the following condition:

For $\theta, \sigma \in X^*$ and $f, g \in C_0(S,X)$, if $U_\theta f = U_\sigma g$, then $\theta(Tf) = \sigma(Tg)$.
$$(\#I)$$

Condition $(\#I)$ in this context, is analogous to condition $(\#C)$ in Definition **3.3.4**.

It is easy to check that $\mathcal{P}$ is a subspace of the space $L\left(C_0\left(S,X\right),X\right)$ of all bounded operators from $C_0\left(S,X\right)$ to X. Also one can prove that $\mathcal{P}$ is closed in the weak operator topology of $L\left(C_0\left(S,X\right),X\right).$ Note also that for a given bounded $T:C_0\left(S,X\right)\to X$, condition $(\#P)$ implies condition $(\#I)$ i.e $T\in\mathcal{P}.$ The crucial point is that condition $(\#I)$ implies condition $(\#P)$ for some bounded scalar measure μ on $\mathcal{B}(S).$ To prove this fact, the following theorem is basic:

Theorem 4.1.7. **[Mez08]** *Let $T:C_0\left(S,X\right)\to X$ be a bounded operator satsfying condition $(\#I).$ Then there exists a unique bounded funcional φ on $C_0\left(S\right)$ such that:*

$$\varphi\circ U_\theta=\theta\circ T,\quad\forall\theta\in X^*.\qquad(\#4.3)$$

Proof: Let h be fixed in $C_0\left(S\right).$ If $0\neq\theta\in X^*$, by **Lemma 4.1.4** there is $f\in C_0\left(S,X\right)$, solution of $U_\theta f=h.$ Define $\varphi\left(h\right)=\theta\left(Tf\right).$ Then φ is well defined since if $U_\theta f=U_\sigma g=h$, for $\theta,\sigma\in X^*$ and $f,g\in C_0\left(S,X\right)$, then $\theta\left(Tf\right)=\sigma\left(Tg\right).$ It is clear that φ is linear. Also $(\#4.3)$ is immediate by construction. It remains to prove that φ is bounded. Let $h\in C_0\left(S\right)$ and let $0\neq\theta\in X^*$; since every solution f of $U_\theta f=h$ works in the definition of φ, we may choose, $f(t)=h(t)x$, with x fixed in X so that $\theta\left(x\right)=1.$ In this case we have (see **Definition 4.1.1**):

$$\widetilde{p}_{\gamma,K}\left(f\right)=\|h\|\,.p_\gamma\left(x\right)\qquad(\#4.4)$$

where $K=supp(h)$, and p_γ is a seminorm on $X.$ By $(\#4.3)$ $\varphi\left(h\right)=\theta\left(Tf\right)$, and, since θ is bounded, there is a constant $M>0$ and a seminorm p_α on X such that: $|\varphi\left(h\right)|=|\theta\left(Tf\right)|\leq M.p_\alpha\left(Tf\right).$ But T is bounded; so for each compact $K\subseteq S$ and for the preceding p_α, there is a constant $\lambda>0$ and a seminorm $\widetilde{p}_{\beta,K}$ on $C(S,K,X)$ such that:

$$p_\alpha\left(Tf\right)\leq\lambda.\widetilde{p}_{\beta,K}\left(f\right).$$

Appealing to $(\#4.4)$, with $\gamma=\beta$, we get:

$$p_\alpha\left(Tf\right)\leq\lambda.\|h\|\,.p_\beta\left(x\right).$$

Now, with the above estimation of $|\varphi\left(h\right)|,$ we deduce that

$$|\varphi\left(h\right)|\leq M.\lambda.p_\beta\left(x\right).\|h\|,$$

which proves the boundedness of $\varphi.$ Uniqueness comes from the fact that U_θ is onto. $\square$

As a consequence we have the main representation theorem:

Theorem 4.1.8. [Mez08] *Let* $T : C_0(S, X) \to X$ *be in the class* $\mathcal{P}$. *Then there is a unique bounded signed measure* μ *on* $\mathcal{B}(S)$ *such that* $(\#P)$ *holds. Moreover for each seminorm* p_α *on* X *we have* $|T|_{p_\alpha} = |\mu|$, *where* $|\mu|$ *is the total variation of* μ *and* $|T|_{p_\alpha}$ *is the* $p_\alpha-$*norm of* T *defined by*

$$|T|_{p_\alpha} = \sup\{p_\alpha(Tf) : f \in \widetilde{B}_{p_\alpha}\}$$

with $\widetilde{B}_{p_\alpha} = \{f \in C_0(S, X) : \|f\|_{p_\alpha, \infty} = \sup_{s \in S} p_\alpha(f(s)) \leq 1\}$.

Proof: Let φ be given by **Theorem 4.1.7**. By the Riesz representation theorem, there exists a unique bounded signed measure μ on $\mathcal{B}(S)$ such that:

$$\varphi(h) = \int_S h(s) \ d\mu(s), \quad \forall h \in C_0(S). \qquad (\#4.5)$$

Taking h of the form $h = U_\theta f = \langle \theta, f(\bullet) \rangle$, with $f \in C_0(S, X)$ and citing $(\#4.3)$ again, yields

$$\varphi \circ U_\theta f = \theta \circ Tf = \int_S \langle \theta, f(s) \rangle \ d\mu(s),$$

which is $(\#\mathbf{P})$. Now to compute $|T|_{p_\alpha}$, observe from the integral form of $\theta \circ Tf$ that

$$|\theta \circ Tf| \leq \sup\{|\langle \theta, f(s) \rangle| : s \in S\} \cdot |\mu|.$$

Taking the supremum in both sides over $\theta \in B^o_{p_\alpha}$, the polar set of the unit ball $B_{p_\alpha} = \{x \in X, \ p_\alpha(x) \leq 1\}$ of X, we get:

$$
\begin{aligned}
\sup_{\theta \in B^o_{p_\alpha}} |\theta \circ Tf| \ = \ p_\alpha(Tf) &\leq \ \sup_{\theta \in B^o_{p_\alpha}} \sup_{s \in S} |\langle \theta, f(s) \rangle| \cdot |\mu| \\
&= \ \sup_{s \in S} \sup_{\theta \in B^o_{p_\alpha}} |\langle \theta, f(s) \rangle| \cdot |\mu| \\
&= \ \sup \ p_\alpha(f(s)) \cdot |\mu| \leq |\mu|, \text{ for } f \in \widetilde{B}_{p_\alpha}.
\end{aligned}
$$

So we deduce that $|T|_{p_\alpha} \leq |\mu|$. To see the reverse inequality, let us consider a function $f \in C_0(S, X)$ of the form $f = g.x$, with $g \in C_0(S)$ satisfying $\|g\| \leq 1$ and x fixed in X such that $p_\alpha(x) = 1$. With this choice, the function f belongs to the unit ball $\widetilde{B}_{p_\alpha}$. Then we have $\langle \theta, f(s) \rangle = g(s).\theta(x)$ and

$$\langle \theta, Tf \rangle = \int_S \langle \theta, f(s) \rangle \ d\mu(s) = \theta(x) \int_S g(s) \ d\mu(s);$$

so that

$$\sup_{\theta \in B^o_{p_\alpha}} |\theta \circ Tf| = p_\alpha(Tf) = p_\alpha(x) \left| \int_S g(s) \ d\mu(s) \right| = \left| \int_S g(s) \ d\mu(s) \right|,$$

since $p_\alpha(x) = 1$. So we get

$$p_\alpha(Tf) = \left| \int_S g(s) \ d\mu(s) \right| \leq |T|_{p_\alpha}$$

because $f \in \widetilde{B}_{p_\alpha}$. Therefore

$$\sup \left\{ \left| \int_S g(s) \, d\mu(s) \right|, \ g \in C_0(S), \ \|g\| \le 1 \right\} = |\mu| \le |T|_{p_\alpha} . \ \square$$

By this theorem we may denote each operator T in the class $\mathcal{P}$ by the conventional symbol

$$Tf = P - \int_S f(s) \, d\mu(s), \ \ f \in C_0(S, X), \tag{#4.6}$$

where the letter P stands for Pettis integral.

Remark 4.1.9.. Usually a weak integral is defined as a vector x^{**} in the second conjugate space X^{**} (see the Dunford integral in [**DU77**]). The construction of the Pettis integral, that is a Dunford integral with values in X, is not so straightforward and needs additional conditions on the space X. In our present setting we were able to construct directly a whole class of Pettis integrals with values in the topological vector space X. In **Section 4.2** below, we shall consider the reverse direction, that is, we start with a bounded measure μ on S and we will construct directly the Pettis integrals under μ by means of a bounded operator T in the class $\mathcal{P}$. But this will be achieved under additional assumptions on X.

4.2 Operators Associated to Scalar Measures via Pettis Integrals

In this section we start with a bounded scalar measure μ on $\mathcal{B}(S)$ and we seek for a linear bounded $T : C_0(S, X) \to X$ such that the correspondence between μ and T would be given by (#**P**). First let us make some observations.

Remark 4.2.1. A little inspection of (#**P**) suggests the following quite plausible observations: First the integral $\int_S \langle \theta, f(s) \rangle \, d\mu(s)$, as a linear functional of θ on X^*, should be at least continuous for some convenient topology on X^*. Also the existence of the corresponding Tf in (#**P**) will require that such topology on X^* should be compatible for the dual pair $\langle X^*, X \rangle$. Finally, to get the continuity of the functional $\theta \to \int_S \langle \theta, f(s) \rangle \, d\mu(s)$, one can seek conditions such that if $\theta \to 0$ in an appropriate manner, then $\langle \theta, f(s) \rangle$ goes to 0 uniformly for $s \in S$. Since μ is bounded this will give $\int_S \langle \theta, f(s) \rangle \, d\mu(s) \to 0$.

In what follows we shall show that such a program can be realized for a locally convex space having the convex compactness property (see **4.2.4** below).

Definition 4.2.2. We shall denote by X^*_τ the dual space X^* equipped with the **Mackey topology** $\tau(X^*, X)$. According to **Mackey-Arens theorem**, it is the topology of uniform convergence on the family of absolutely convex

$w(X, X^*)-$compact sets of X. Also it is the largest compatible topology for the dual pair $\langle X^*, X \rangle$. Then we have the well known:

Proposition 4.2.3. *For each* $x^{**} \in (X_\tau^*)^*$ *there exists a unique* $x \in X$ *such that* $x^{**}(\theta) = \theta(x)$, $\forall \theta \in X^*$.

Definition 4.2.4. A locally convex space X is said to have the **convex compactness property** if for every compact set $K \subseteq X$, the absolute convex closure K_0 of K is also compact. For example, every quasicomplete locally convex space has the convex compactness property.

Theorem 4.2.5. **[Mez08]** *Let* X *be a locally convex space with the convex compactness property, and whose dual* X^* *is equipped with the Mackey topology* $\tau(X^*, X)$. *If* μ *is a bounded scalar measure on* $\mathcal{B}(S)$, *then there is a unique bounded operator* $T : C_0(S, X) \to X$ *in the class* $\mathcal{P}$ *satisfying* $(\#P)$ *of* **Definition 4.1.5**, *with* $|T|_{p_\alpha} = |\mu|$ *for each seminorm* p_α *on* X.

Proof: Fix f in $C_0(S, X)$ and define the functional $\Gamma_f : X^* \to \mathbb{R}$, by

$$\Gamma_f(\theta) = \int_S \langle \theta, f(s) \rangle \, d\mu(s).$$

It is clear that Γ_f is linear. Moreover $\Gamma_f \in (X_\tau^*)^*$. Indeed it is enough to prove that $\lim_{\theta \to 0} \Gamma_f(\theta) = 0$. If $\theta \to 0$, in X_τ^*, then for each absolutely convex $w(X, X^*)$-compact $B \subseteq X$, $\theta(x) \to 0$ uniformly on B. But since $f \in C_0(S, X)$, the set $K = \{f(s) : s \in S\}$ is compact.

Therefore, by the convex compactness property we deduce that K_0, the absolutely convex closure of K is compact, hence weakly compact; so $\theta \to 0$ uniformly on K_0. Consequently $\langle \theta, f(s) \rangle \to 0$ uniformly in $s \in S$.

Hence $\Gamma_f(\theta) = \int_S \langle \theta, f(s) \rangle \, d\mu(s) \to 0$, because μ is bounded. We deduce that $\Gamma_f \in (X_\tau^*)^*$; by **Proposition 4.2.3**, there is a unique $\xi_f \in X$ such that $\Gamma_f(\theta) = \langle \theta, \xi_f \rangle$, $\forall \theta \in X^*$. Now let us define the operator $T : C_0(S, X) \to X$, by

$$Tf = \xi_f, f \in C_0(S, X).$$

It is easily checked that T is linear, and satisfies $(\#P)$ of **Definition 4.1.5** by construction. We have to show that T is bounded. Let p_α be a seminorm on X, and let K be a compact subset of X. For $f \in C(S, K, X)$, we have:

$$
\begin{aligned}
p_\alpha(\xi_f) \quad &= \quad p_\alpha(Tf) = \sup_{\theta \in B_{p_\alpha}^o} |\theta \circ Tf| = \sup_{\theta \in B_{p_\alpha}^o} \left| \int_S \langle \theta, f(s) \rangle \, d\mu(s) \right| \\
&\leq \quad \sup_{\theta \in B_{p_\alpha}^o} \sup_{s \in K} |\langle \theta, f(s) \rangle| \cdot |\mu| \\
&= \quad \sup_{s \in K} \sup_{\theta \in B_{p_\alpha}^o} |\langle \theta, f(s) \rangle| \cdot |\mu| = \widetilde{p}_{\alpha, K}(f) \cdot |\mu|
\end{aligned}
$$

which proves the continuity of T. The relation $|T|_{p_\alpha} = |\mu|$ is proved as in Theorem **4.1.8**. $\square$

Theorems **4.1.8**, and **4.2.5** may be put together to give:

Theorem 4.2.6. **[Mez08]** *If* X *is a locally convex space having the convex compactness property, then there is an isometric isomorphism between* *the space $\mathcal{P}$ and the topological dual $C_0^*(S)$ of the space $C_0(S)$. In this isomorphism the operator $T \in \mathcal{P}$ corresponds to the measure $\mu \in C_0^*(S)$ via the integral representation:*

$$\langle \theta, Tf \rangle = \int_S \langle \theta, f \rangle \, d\mu, \qquad |T|_{p_\alpha} = |\mu|, \forall \theta \in X^*, \ \forall f \in C_0(S, X).$$

Remark 4.2.7. One essential point in the proof of **Theorem 4.2.5** was the uniform convergence in $s \in S$ of $\langle \theta, f(s) \rangle$ to 0 when $\theta \to 0$ inX_τ^*. This has given $\Gamma_f \in (X_\tau^*)^*$. If we want to get rid of the convex compactness condition, we must have $\Gamma_f(\theta) \to 0$ when $\theta \to 0$ inX_τ^*. But if $\theta \to 0$ inX_τ^*, we certainly have $\langle \theta, f(s) \rangle \to 0$ for each $s \in S$. Then, as the set $\{\langle \theta, f(s) \rangle : s \in S\}$ is bounded for each $\theta \in X^*$, getting uniform convergence with respect to s is reminiscent to a uniform boundedness principle, which in the present setting, should be valid for the dual X_τ^* of X. It is well known that a general version of this principle has been stated, via equicontinuity, for the barrelled topological vector spaces, [**Wila78**], Chapter **9**, **Theorem 9.3.4**). With this observation in mind we can state:

Theorem 4.2.8. [**Mez08**] *Let X be a locally convex Hausdorff space whose dual X_τ^* is a barrelled space. If μ is a bounded signed measure on $\mathcal{B}(S)$, then there is a unique bounded operator $T : C_0(S, X) \to X$ in the class $\mathcal{P}$ satisfying $(\#P)$ with respect to μ and such that $|T|_{p_\alpha} = |\mu|$.*

Proof: Consider the set of the linear functionals on X_τ^*, $\mathcal{F} = \{\langle \bullet, f(s) \rangle, s \in S\}$. Since for each θ the function $s \to \langle \theta, f(s) \rangle$ is continuous and since $f \in C_0(S, X)$, we deduce that for each $\theta \in X_\tau^*$ the set $\{\langle \theta, f(s) \rangle, s \in S\}$ is bounded, that, is the family $\mathcal{F}$ is pointwise bounded. Since X_τ^* is barrelled, the family $\mathcal{F}$ is equicontinuous, by the uniform boundedness principle.Therefore if $\theta \to 0$ inX_τ^*, for each $\varepsilon > 0$ there is a neighborhood V of 0 in X_τ^* such that if $\theta \in V$ we have $|\langle \theta, f(s) \rangle| \leq \varepsilon$, for all $s \in S$. This means that $\langle \theta, f(s) \rangle \to 0$ uniformly in $s \in S$. This gives the continuity of the functional

$$\Gamma_f(\theta) = \int_S \langle \theta, f(s) \rangle \, d\mu(s).$$

Now the proof goes along the same lines as in the proof of **Theorem 4.2.5**. $\square$

Since a locally convex space, which is a Baire space, is a barrelled space, we deduce that **Theorem 4.2.8** is valid for a space whose dual X_τ^* is a Baire space.

Most of the results in this section have been obtained for a space whose dual is a Mackey space. It is natural to ask if similar representations can be established if the dual is endowed with another topology, e.g. the strong topology. In what follows, a construction similar to that of **Theorem 4.2.5** can performed for semireflexive spaces.

Definition 4.2.9. The **strong topology** $\beta(X^*, X)$ of X^* is the topology generated by the family of the seminorms:

$$p_B(\theta) = \sup_{x \in B} |\theta(x)|, \qquad\qquad \theta \in X^*, \qquad\qquad (\#4.7)$$

B running over all bounded sets of X.

It is the topology of uniform convergence on the bounded sets of X. When we restrict $(\#4.7)$ to the finite sets B of X we get the so called weak* topology $w\,(X^*, X)$, which is the topology of simple convergence on X. We shall denote by $X_b^*\,(X_w^*)$ the space X^* equipped with the $b\,(X^*, X)-$topology (the $w\,(X^*, X)-$topology).Then we have:

Proposition 4.2.10. (a) *For each $x^{**} \in (X_w^*)^*$ there exists a unique $x \in X$ such that: $x^{**}\,(\theta) = \theta\,(x)\,, \forall \theta \in X_w^*$.*

(b) *$(X_\sigma^*)^* \subseteq \left(X_\beta^*\right)^*$, that is, every weak* continuous functional on X^* is strongly continuous.*

Definition 4.2.11. *We say that the space X is semireflexive if $(X_w^*)^* = (X_b^*)^*$.*

The following theorem is, as far as we know, new.

Theorem 4.2.12. *Let X be a locally convex Hausdorff semireflexive space. If μ is a bounded signed measure on $\mathcal{B}(S)$, then there is a unique bounded operator $T : C_0\,(S, X) \to X$ in the class $\mathcal{P}$ satisfying:*

$$\begin{aligned}
\langle \theta, Tf \rangle &= \int_S \langle \theta, f \rangle\, d\mu,\ \forall f \in C_0\,(S, X), &&(\#4.8)\\
|T|_{p_\alpha} &= |\mu|.
\end{aligned}$$

Proof: Fix f in $C_0\,(S, X)$ and define the functional $\Lambda_f : X^* \to \mathbb{R}$, by

$$\Lambda_f\,(\theta) = \int_S \langle \theta, f\,(s) \rangle\ d\mu\,(s).$$

It is clear that Λ_f is linear. Moreover $\Lambda_f \in (X_b^*)^*$. Indeed it is enough to prove that $\lim_{\theta \to 0} \Lambda_f\,(\theta) = 0$. If $\theta \to 0$, in X_b^*, then for each bounded set $B \subseteq X$, $\theta\,(x) \to 0$ uniformly for $x \in B$. But since $f \in C_0\,(S, X)$, the set $\{f\,(s) : s \in S\}$ is bounded, so $\langle \theta, f\,(s) \rangle \to 0$ uniformly in $s \in S$. Therefore.

$$\int_S \langle \theta, f\,(s) \rangle\ d\mu\,(s) \to 0,$$

because the measure μ is of bounded variation. Hence $\Lambda_f \in (X_b^*)^*$. Since X is semireflexive (**Section A.4**), Λ_f is in $(X_\sigma^*)^*$; by **Proposition 4.2.10(a)**, there is a unique $\zeta_f \in X$ such that

$$\Lambda_f\,(\theta) = \langle \theta, \zeta_f \rangle, \forall \theta \in X^*.$$

Now let us define the operator $T : C_0\,(S, X) \to X$, by

$$Tf = \zeta_f,\quad f \in C_0\,(S, X).$$

It is easily checked that T is linear, and satisfies $(\#4.8)$ by constuction. We have to show that T is bounded. Let p_α be a seminorm on X, and let K

be á compact subset of X. For $f \in C(S, K, X)$, we have:

$$
\begin{aligned}
p_\alpha\left(\zeta_f\right) &= p_\alpha\left(Tf\right) = \sup_{\theta \in B^o_{p_\alpha}} |\theta \circ Tf| = \sup_{\theta \in B^o_{p_\alpha}} \left|\int_S \langle \theta, f(s)\rangle \, d\mu(s)\right| \\
&\leq \sup_{\theta \in B^o_{p_\alpha}} \sup_{s \in K} |\langle \theta, f(s)\rangle| \cdot |\mu| \\
&= \sup_{s \in K} \sup_{\theta \in B^o_{p_\alpha}} |\langle \theta, f(s)\rangle| \cdot |\mu| = \widetilde{p}_{\alpha,K}(f) \cdot |\mu| \; PART
\end{aligned}
$$

which proves the continuity of T. The relation $|T|_{p_\alpha} = |\mu|$ is proved as in **Theorem 4.1.8**.

4.3   Notes and comments

Usually a weak integral is defined as a vector x^{**} in the second conjugate space X^{**} (see the **Dunford integral** in [**DU77**]). A Pettis integral, is a Dunford integral with values in X. Its construction is not so straightforward and needs additional conditions on the space X. In our present setting we were able to construct directly a whole class of Pettis integrals representing bounded operators with values in the topological vector space X, which are the operators in the class $\mathcal{P}$. In **Section 4.2** we considered the reverse direction, that is, we start with a bounded measure μ on S and we will construct directly the Pettis integrals under μ by means of a bounded operator T in the class $\mathcal{P}$. But this has been achieved under the additional assumption of convex compactness property on X (**Theorem 4.2.6**). The same result is obtained for a locally convex TVS whose Mackey dual X^*_τ is a barrelled space (**Theorem 4.2.8**) and for a semireflexive space (**Theorem 4.2.12**), see [**Mez08**], for details. One may ask along these lines about the identificaion of the most general space for which the representation of **Theorem 4.2.6** holds.

5

Representations of Weakly Compact Operators in LCSs

In this Chapter, we consider two extensions of the representations given in **Sections 3.1** and **3.2**:

The first one, due to **D.R. Lewis [Lew70]**, develops an appropriate integration process with respect to measures into a locally convex Hausdorff space X, and gives integral representation for a weakly compact operator $T : C(S) \longrightarrow X$, with respect to an $X-$valued measure in the case of a compact Hausdorff S. This extends the representation of Bartle-Dunford-Schwartz given in **Theorem 3.1.5** given for a Banach space X.

The second representation is similar to that of **Dinculeanu-Singer** given in **Theorem 3.2.1**. It is due to **A. Katsaras** and **D. B. Liu [KL 75]**, who considered an entirely different setting from that used by Dinculeanu-Singer. They have been able to get a representation for weakly compact operators, with respect to an operator valued measure for some special function spaces (see **Section 5.3** below for details).

5.1   Integration in Locally Convex Spaces

In this section we summarize the main ingredients of the integration process which will be used. Let $(S, \mathcal{B})$ be a measurable space and let X be a locally convex Hausdorff space with dual X^*. We denote by μ a $\sigma-$additive measure on $\mathcal{B}$ with values into X. For $x^* \in X^*$ and p a semi-norm on X, the symbol $x^* \leq p$ means $|x^*(x)| \leq p(x)$ for all $x \in X$.

Definition 5.1.1: If p is a semi-norm on X, then the $p-$**semivariation** of μ is the function from $\mathcal{B}$ into extended reals defined by:

$$\|\mu\|_p (X) = \underset{x^* \leq p}{Sup}\, v(x^*\mu, X)$$

where $v(x^*\mu,)$ is the scalar variation of $x^*\mu$.

It is not difficult to check that $\|\mu\|_p(\cdot)$ is monotone, countably subadditive and real valued. Moreover we have for each $E \in \mathcal{B}$:

$$p(\mu(E)) \leq \|\mu\|_p(E) \leq 4\underset{Y \subseteq E}{Sup}\, p(\mu(Y))$$

Definition 5.1.2: A function $f : S \longrightarrow \mathbb{R}$ is μ-**integrable** if:
(i) f is $x^*\mu$-integrable for each $x^* \in X^*$,

(ii) for each $E \in \mathcal{B}$ there is an element of X denoted by $\int_E f(t)\,\mu(dt)$ such that:

$$x^* \int_E f(t)\,\mu(dt) = \int_E f(t)\,x^*\mu(dt).$$

Remarks. It is easy to see that the integral has the following properties:
(1) The integral is linear.
(2) Every simple function $\sum_{i \leq n} a_i \chi_{E_i}$ is μ-integrable and

$$\int_E \left(\sum_{i \leq n} a_i \chi_{E_i} \right)(t)\,\mu(dt) = \sum_{i \leq n} a_i \mu\left(E \cap E_i \right).$$

(3) If f is bounded and μ-integrable, then $s \in E$ for each $E \in \mathcal{B}$ and continuous semi-norm p.

(4) If f is μ-integrable and T is a continuous linear operator from X into a locally convex Hausdorff space Y, then f is $T\mu$-integrable
and

$$\int_E f(t)\,T\mu(dt) = T \int_E f(t)\,\mu(dt), \quad E \in \mathcal{B}.$$

The following theorem gives a criterion for μ-integrability in the case of a sequentially complete space X. see [Lew70]:

Theorem 5.1.3: *Suppose X is sequentially complete and f is a real valued function on S. The following are equivalent:*

(1) f is μ-integrable.

(2) There is a sequence (f_n) of bounded measurable functions which converges pointwise to f and for which $\left(\int_E f_n(t)\,\mu(dt) \right)$ *is Cauchy uniformly with respect to $E \in \mathcal{B}$.*

(3) There is a sequence (f_n) of simple functions which converges pointwise to f and for which $\left(\int_E f_n(t)\,\mu(dt) \right)$ *is Cauchy for each $E \in \mathcal{B}$.*

5.2   Representation of Weakly Compact Operators

Below S is a compact Hausdorff space, $\mathcal{B}o(S)$ the Borel sets of S, $C(S)$ the Banach space (under supremum norm) of continuous complex valued functions on S and X a locally convex, Hausdorff space.

Theorem 5.2.1. *Let $A : C(S) \to X$ be a weakly compact linear operator. There is a measure $\mu : \mathcal{B}o(S) \to X$ such that*

(1) μ *is regular,*

(2) *the closed absolutely convex hull of* $\mu[\mathcal{B}o(S)]$ *is weakly compact,*

(3) *every bounded Borel function on* S *is* μ-*integrable,*

(4) $Af = \int_S f(t)\mu(dt)$ *for* $f \in C(S)$, *and*

(5) $A^*x^* = x^*\mu$ *for* $x^* \in X^*$.

Remarks. **(a)** *Conditions* **(1)** *and* **(4)** *define* μ *uniquely.*

(b) $\|A\| = \|\mu\|(S)$ *whenever* X *is normed.*

(b) *If* μ *is a measure on* $\mathcal{B}o(S)$ *which satisfies* **(1)**, **(2)**, *and* **(3)**, *then* **(4)** *defines a weakly compact operator which satisfies* **(5)**.

(c) *If* X *is complete,then* **(2)** *and* **(3)** *follow from* **(1)**.

Proof. Since the dual of $C(S)$ may be identified with the bounded regular Borel measures on Bo(S), the equation

$$\overline{g}(\lambda) = \int_S g(t)\lambda(dt).$$

defines an element of $C(S)^{**}$ for each bounded Borel function g. Since A is weakly compact, A^{**}, the algebraic adjoint of A^*, maps $C(S)^{**}$ into X. For $E \in$ Bo(S), let $\mu(E) = A^{**}(\overline{\chi_E})$. For each $x^* \in X^*$, $x^*\mu = A^*x^*$ is a regular measure, so μ is a regular measure. Since A^{**} maps the unit ball of $C(S)^{**}$ into a weakly compact subset of X, condition (3) is satisfied. If $\sum_{1 \le i \le n} a_i \chi_{E_i}$ is a simple Borel function, then

$$A^{**}\left(\overline{\sum_{1 \le i \le n} a_i \chi_{E_i}}\right) = \sum_{1 \le i \le n} a_i A^{**}\left(\overline{\chi_{E_i}}\right) = \int_E \left(\sum_{1 \le i \le n} a_i \chi_{E_i}\right)(t)\,\mu(dt)$$

Thus $x^* A^{**}(\overline{g}) = \int_S g(t) x^*\mu(dt)$ holds for each bounded Borel function g and $,x^* \in X^*$. Finally, $\|A\| = \|A^*\| = \|\mu\|(E)$ if X is normed.

Conversely, suppose μ satisfies **(1)**, **(2)**, **(3)**. The operator A defined by **(4)** is continuous since $p[Af] \le \|f\|\,\|\mu\|_p(S)$ for each continuous semi-norm p on X. Also, by the regularity of μ, $A^*x^* = x^*\mu$ for $x^* \in X^*$. Let U be the polar of the closed, absolutely convex hull of $\mu[\text{Bo}(S)]$. U is a neighborhood of zero in the Mackey topology on X^*, and, for $x^* \in U$, $\|A^*x^*\| \le 4$. Thus A^* is continuous with the Mackey topology on X^* and the norm topology on $C(S)^*$-this implies that A is weakly compact. If X is complete and μ is regular, then

$$Af = \int_S f(t)\mu(dt)$$

defines a continuous linear operator for $C(S)$ into X such that $A^*x^* = x^*\mu$ for $x^* \in X^*$.

To see that **(2)** holds, it is sufficient to show that A is weakly compact-equivalently, that A^* maps equicontinuous sets into weakly relatively compact sets. Let V be an open neighborhood of zero in X generated by a semi-norm p. Let V^o be the polar set of V. For $x^* \in V^o$ and $E \in$ Bo(S)

$$|x^*\mu(E)| \le \|\mu\|_p(E),$$

so the countable additivity of $A^*[V^o]$ is uniform. This together with norm boundedness implies that $A^*[V^o]$ is relatively weakly compact. $\square$

Remarks. (1) In [**BDS**55], **Bartle, Dunford and Schwartz** have given a similar integral representation for weakly compact operators from $C(S)$ into a Banach space.

(2) **Grothendieck** [**Gro**53] has noted that there is a one-to-one correspondence between the weakly compact operators from $C(S)$ into a complete locally convex space X the $X-$valued measures on the Baire sets of S, although he did not give an integral representation of such operators.

(3) Finally we note that **Theorem 5.2.1** may be generalized in the following way:

Let T be a locally compact Hausdorff space. $C_0(T)$ (resp. $C_c(T)$) is the Banach space under the sup. norm of scalar functions on T which vanish at infinity (resp. have compact support).$C(T)_\beta$ is the space of bounded continuous scalar functions on T topologized by the the family of seminorms

$$p_\varphi\left(f\right) = \sup_{t \in T} |\varphi\left(t\right) f\left(t\right)|, \text{ where } \varphi \in C_0(T).$$

This is the strict topology introduced by **R. C. Buck** in [**Buc**58], in the more general setting of locally convex spaces. A weakly compact operator $A : C_0(T) \to X$ can be represented by integration with respect to $X-$valued measure since A can be extended to a weakly compact operator on the space of continuous functions over the one point compactification of T. Weakly compact operators on $C_c(T)$ have such a representationsince they can be extended to $C_0(T)$. The bounded sets of $C(T)_\beta$ are precisely the uniformly bounded sets and each element of $C(T)_\beta^*$ can be identified with a bounded regular Borel measure on T, [**Buc**58].

5.3   Integration with Operator Valued Measure

Let S be a completely regular Hausdorff space and $\mathcal{B} = \mathcal{B}(S)$ be the algebra of S generated by the zero sets. By $\mathcal{B}a = \mathcal{B}a(S)$ and $\mathcal{B}o = \mathcal{B}o(S)$ we will denote the $\sigma-$algebras of Baire and Borel sets respectively. Let $M(\mathcal{B})$ be the space of all bounded additive regular (with respect to zero sets) measures on $\mathcal{B}$.

Definition 5.3.1: Let X be a real locally convex Hausdorff space. For p a continuous seminorm on X, we define $M_p(\mathcal{B}, X^*)$ as the set of all X^*-valued (X^* is the dual of X) finitely-additive measures μ on $\mathcal{B}$ with the following two properties:

(1) For every $x \in X$, the function μx, from $\mathcal{B}$ into the reals R, $G \to \mu(G)x$, is in $M(\mathcal{B})$.

(2) $\|\mu\|_p = \mu_p(S) < \infty$,

where for G in $\mathcal{B}$ the $\mu_p(G)$ is defined to be the supremum of all $|\sum \mu(G_i)x_i|$ for all finite $\mathcal{B}$-partitions $\{G_i\}$ of G, i.e., $G_i \in \mathcal{B}$, and all finite collections

$x_i \in B_p = \{x \in X : p(x) \leqq 1\}$.

Notation. Denote by $C_{rc}(S, X)$ the space of all continuous functions f from S into X for which $f(S)$ is relatively compact. Every f in $C_{rc}(S, X)$ has a unique continuous extension $\hat{f}$ to all of the **Stone-Čech compactification** βS. By $C_b(S)$ we denote the space of all bounded continuous real-valued functions on S.

For any $p \in I$, we denote the p-unit ball

$$B_p = \{x \in X : p(x) \leq 1\}$$

and its polar by

$$B_p^o = \{x^* \in X^* : \sup_{x \in B_p} x^*(x) \leq 1\}.$$

Definition 5.3.2: For G in $\mathcal{B}$ and $\mu \in M_p(\mathcal{B}, X^*)$, if $f \in C_{rc}(S, X)$, we define the **integral** $\int_G f d\mu$ by

$$\int_G f d\mu = \lim \Sigma \mu(G_i) f(x_i)$$

where the limit is taken over the directed set of all finite $\mathcal{B}$-partitions $\{G_i\}$ of G and $x_i \in G_i$.

Remarks. (1) The map $f \to T_\mu(f) = \int_X f d\mu$ is a uniformly continuous linear functional on $C_{rc}(S, X)$.

(2) Moreover, $\|\mu\|_p = \sup\{|T_\mu(f)| : \|f\|_p \leqq 1\}$.

(3) The mapping $\mu \to T_\mu$ is a one-to-one linear map from $M(\mathcal{B}, X^*)$ into $(C_{rc}(S, X), u)^*$.

Definition 5.3.3: Let now Y be another real locally convex Hausdorff space and let $\{q : q \in J\}$ be a generating directed family of continuous seminorms on Y. Let $L(X, Y)$ denote the space of all continuous operators from X into Y.

(1) We define $M(\mathcal{B}(S), L(X, Y))$ to be the space of all finitely-additive $L(X, Y)$ valued measures μ on $\mathcal{B}(S)$ with the following two properties:

(a) For each $y^* \in Y^*$ the set function $x^*\mu : \mathcal{B}(S) \to X^*$, $(x^*\mu)(G)x = x^*(\mu(G)x)$, $x \in X$, is in $M(\mathcal{B}(S, X^*)$.

(b) Given $q \in J$ there exists p in I such that for all x^* in the polar B_q^o of B_q in Y, $x^*\mu \in M_p(\mathcal{B}(S), X^*)$ and

$$\|\mu\|_{p,q} = \mu_{p,q} = \mu_{p,q}(X) < \infty,$$

where for Q in $\mathcal{B}(S)$,

$$\mu_{p,q}(Q) = \sup\{(x^*\mu)_p(Q) : x^* \in B_q^o\}.$$

(2) Let $\mu \in M(\mathcal{B}(S), L(X, Y))$ and f a function from X into Y. We say that f is μ**-integrable** over G in $\mathcal{B}(S)$ if

(i) For each $y^* \in Y^*$, the integral $\int_G f d(x^*\mu)$ exists.

(ii) there exists a vector in Y denoted by $\int_G f d\mu$ such that for all $y^* \in Y^*$ we have

$$x^*\left(\int_G f d\mu\right) = \int_G f d(x^*\mu).$$

Remarks. (I) Since Y is a locally convex Hausdorff space, the $\int_G f d\mu$ is unique whenever it exists.

(II) If f is μ-integrable over all G in $\mathcal{B}(S)$, we say that f is μ-integrable.

5.4   Linear Operators from $C_{rc}(S, X)$ into F

Let $X, Y, \{p : p \in l\}, \{q : q \in J\}$ be as in **Section 5.3**. Recall that a linear operator T from a topological vector space A into another B is weakly compact if it maps bounded subsets of A into weakly relatively compact subsets of B. We need the following lemma due to **Grothendieck** .

Lemma 5.4.1. *Let T be an operator from a topological vector space A into another B and let T^* and T^{**} denote, respectively, the transpose and the second transpose of T. The following are equivalent:*

(1) T is weakly compact.

(2) T^{**} maps A^{**} into B.

(3) If B^* is equipped with the Mackey topology $m(B^*, B)$ and A^* with the strong topology $b(A^*, A)$, then T^* is continuous.

Lemma 5.4.2. *Let $g \in C_{rc}(S, X)$ and G in $\mathcal{B}(S)$. Define ϕ on $M(\mathcal{B}(S), X^*)$ by*

$$\phi(\mu) = \int_G g d\mu, \ \mu \in M(\mathcal{B}(S), X^*).$$

Then ϕ belongs to the $(C_{rc}(S, X), u)^{**}$.

Proof Let

$$A = \{f \in C_{rc}(S, X) : ||f||_p \leqq ||g||_p \text{ for all } p \in I\}.$$

Then A is u-bounded in $C_{rc}(S, X)$ and hence the polar A^o in $(C_{rc}(S, X), u)^*$ is a strong neighborhood of zero. We will finish the proof by showing that ϕ is bounded on A°. To this end consider an arbitrary μ in A°. Let $\varepsilon > 0$ be given. There exists a $\mathcal{B}(S)$-partition $G_1, G_2..., G_n$ of G and $x_i \in G_i$ such that

$$\left|\int_G g d\mu\right| \leqq |\Sigma\mu(G_i)x_i| + \varepsilon, x_i = g(x_i).$$

By the regularity of μx_i we can find zero sets $Z_i \subseteq G_i$ such that

$$|\Sigma\mu(G_i)x_i| \ \leqq |\Sigma\mu(Z_i)x_i| + \varepsilon.$$

Again by the regularity of $|\mu x_i|$ ($|\mu x_i|$ is the absolute variation of μx_i) we can find pairwise disjoint cozero sets $U_1, ..., U_n$, $Z_i \subseteq U_i$ such that

$$\sum |\mu x_i| (Z_i - U_i) < \varepsilon.$$

For each $1 \leq i \leq n$, choose $h_i \in C_b(S)$, with $0 \leq h_i \leq 1$, such that $h_i = 1$ on Z_i and $h_i = 0$ on $S - U_i$. Set $h = \Sigma h_i x_i$. Then $h \in A$ and hence $\left| \int_S h d\mu \right| \leq 1$. But

$$\left| \int_S h d\mu \right| \geq \left| \sum \int_{Z_i} h_i x_i d\mu \right| - \left| \sum \int_{U_i - Z_i} h_i d_i \mu x_i \right|$$

$$\geq \left| \sum \mu(Z_i) x_i \right| - \varepsilon \geq \left| \int_S g d\mu \right| - 3\varepsilon.$$

Since $\varepsilon > 0$ was arbitrary we conclude that $\left| \int_G g d\mu \right| \leq 1$. $\square$

Theorem 5.4.3. *Let $T : C_{rc}(S, X) \to Y$ be a continuous weakly compact operator. Then there exists a unique $\mu \in M(B(S), L(X, Y))$ such that:*

(1) *Every $f \in C_{rc}(S, X)$ is μ-integrable and $\int_S f d\mu = T(f)$.*

(2) *If $p \in I$ and $q \in J$ are such that $\|T\|_{p,q} = \sup\{q(T(f)) : \|f\|_p \leq 1\} < \infty$, then $\|\mu\|_{p,q} = \|T\|_{p,q}$.*

(3) *For every $y^* \in Y^*$, we have $T^* y^* = y^* \mu$, so $y^* \mu \in M(B(S), X^*)$.*

(4) *For every bounded set $H \subseteq X$ the set*

$$V_{\mu,H} = \{\Sigma \mu(G_i)(x_i) : \{G_i\} \text{ is a finite } \mathcal{B}(S)\text{-partition of } S, \, x_i \in H\} \subseteq Y$$

is weakly relatively compact.

Conversely, if $\mu \in M(B(S), L(X, Y))$ is such that (4) holds, then every f in $C_{rc}(S, X)$ is μ-integrable and the operator $T : C_{rc}(S, X) \to Y$ given by

$$T(f) = \int_S f d\mu, \, f \in C_{rc}(S, X),$$

is u-continuous and weakly compact.

Proof. Existence of μ: Suppose that $T : C_{rc}(S, X) \to Y$ is u-continuous and weakly compact. By **Lemma 5.4.1**, T^{**} maps $(C_{rc}(S, X), u)^{**}$ into Y. If $f \in C_{rc}(S, X)$ and G in $B(S)$, the function $f \chi_G$ (χ_G is the characteristic function of G) defines an element of $(C_{rc}(S, X), u)^{**}$ by

$$\langle \mu, f \chi_G \rangle = \int_G f d\mu, \, \mu \in M(B(S), X^*) = (C_{rc}(S, X), u)^*.$$

Thus we may consider $f \chi_G$ as an element of $(C_{rc}(S, X), u)^{**}$. Define $\mu(G) : X \to Y$ by

$$\mu(G)x = T^{**}(\chi_G \otimes x), G \text{ in } B(S).$$

It is easy to see that $\mu(G) \in L(X, Y)$. In this way we define a map $\mu : B(S) \to L(X, Y)$ which is clearly finitely additive. If $y^* \in Y^*$ and x in X, then, since $T^* y^* \in X^*$,

$$(y^* \mu)(G)x = y^*(T^{**}(\chi_G \otimes x)) = \langle T^* y^*, \chi_G \otimes x \rangle = T^* y^*(G)x.$$

Thus $y^* \mu = T^* y^* \in M(B(S), X^*)$. Let $q \in J$. Since T is u-continuous there exists $p \in I$ such that $\|T\|_{p,q} < \infty$. Let $y^* \in B_q^o$, where

$$B_q^o = \{y^* \in Y^* : \sup_{y \in B_q} | < y, y^* > | \leq 1\}, \text{ the polar of } B_q \text{ in } Y^*.$$

Then for any $f \in C_{rc}(S, X)$ with $\|f\|_p \leq 1$ we have

$$|\langle f, y^* \mu \rangle| = |\langle f, T^* y^* \rangle| \leq |\langle Tf, y^* \rangle| \leq \|T\|_{p,q} .$$

Thus $\|y^* \mu\|_p \leq \|T\|_{p,q}$ which proves that $\|\mu\|_{p,q} \leq \|T\|_{p,q}$ and so μ is in $M(B(S), L(X, Y))$.

(1) Let $G \in B(S)$ and $f \in C_{rc}(S, X)$. For $y^* \in Y^*$, we have

$$y^*(T^{**}(\chi_G f)) = \langle T^* y^*, \chi_G f \rangle = \langle y^* \mu, \chi_G f \rangle = \int_G f d(y^* \mu).$$

This shows that $\int_G f d\mu = (T^{**}(\chi_G f) \in Y$. Taking $G = S$, we get

$$\int_G f d\mu = T^{**}(f) = T(f).$$

For the uniqueness of μ, suppose μ_1 is another element in $M(B(S), L(X, Y))$ such that $\int_S f d\mu = T(f)$ for all $f \in C_{rc}(S, X)$. Then, for $y^* \in Y^*$, we have

$$\int_S f d(y^* \mu) = \int_S f d(y^* \mu_1) \text{ for all } f \in C_{rc}(S, X).$$

This implies that $y^* \mu = y^* \mu_1$ and hence $\mu = \mu_1$ since Y is a locally convex Hausdorff space.

(2) As proved above, $\|\mu\|_{p,q} \leq \|T\|_{p,q}$. For the reverse inequality, let $f \in C_{rc}(S, X)$ with $\|f\|_p \leq 1$ and $x^* \in B_q^o$. Then we have

$$\left| x^*(T(f)) = \int_S f d(x^* \mu) \right| \leq \|x^* \mu\|_p \leq \|\mu\|_{p,q} .$$

This proves that $\|T\|_{p,q} \leq \|\mu\|_{p,q}$.

(3) This is already proved in part "Existence of μ".

(4) Finally, let H be a bounded subset of X and $W = V_{m,H}$. Let $A = \{f \in C_{rc}(S, X) : f(S) \subseteq H\}$. Then A is u-bounded and therefore $T(A)$ is weakly relatively compact. We will finish the proof of **(4)** by showing that X is contained in the weak closure of $T(A)$. Let $G_1, ..., G_n$ be a $B(S)$-partition of S and $x_1, ..., x_n \in H$. Let $y_1^*, ..., y_N^* \in Y^*$. There exist $q \in J$ and $M > 0$ such that $y_j^* \in MB_q^o$. Let $p \in I$ be such that $\|T\|_{p,q} < \infty$. Since $H \subseteq X$ is bounded, $d = \sup\{p(x) : x \in H\} < \infty$. By the regularity of $(y_j^* \mu)_p$ we can find zero sets $Z_1, ..., Z_n$ with

$$\Sigma_{i=1}^n (y_j^* \mu)_p(G_i - Zi) < \varepsilon/2d$$

(where $\varepsilon > 0$ is arbitrary) for $j = 1, ..., n$. Next, again by regularity of $(y_j^* \mu)_p$, we can find pairwise disjoint cozero sets $U_1, ..., U_n$ with $Z_i \subseteq U_i$ such that for each $j, 1 \leq j \leq N$, we have

$$\Sigma_{i=1}^n (y_j^* \mu)_p(U_i - Zi) < \varepsilon/2d.$$

For each i, $1 \leq i \leq n$, we pick a function $h_i \in C_b(S)$ mith $0 \leq h_i \leq 1$, such that $h_i = 1$ on Z_i and $h_i = 0$ on the complement of U_i. The function $h = \Sigma_1^n h_i \otimes x_i \in A$ and hence $T(h) \in T(A)$. Moreover

$$\left| y_j^*(T(h)) - \sum \mu(G_i)x_i \right|$$
$$= \left| y_j^* \left(\sum \mu(Z_i)x_i - \sum \mu(G_i)x_i + \sum \int_{U_i Z_i} h_i \otimes x_i d\mu \right) \right|$$
$$< \varepsilon/2 + \varepsilon/2 = \varepsilon.$$

This shows that $\sum \mu(G_i)x_i$ is in the weak closure of $T(A)$ and the proof of **(4)** is complete.

Conversely, suppose that $\mu \in M(B(S), L(X, Y))$ satisfies **(4)**. Let $G \in B(S)$ and $f \in C_{rc}(S, X)$. Denote by D_G the set of all $\alpha = \{G_i, ..., G_n; x_i, ..., x_n\}$ where $\{G_i\}$ is a $B(S)$-partition of G and $x_i \in G_i$. For α, γ in D_G we write $\alpha \geq \gamma$ if the $B(S)$ -partition of G for α is arefinement of the one in γ. Then D_G becomes a directed set.

For $\alpha = \{G_1, ..., G_n; x_1, ..., x_n\}$ in D_G, we define $z_\alpha = \sum \mu(G_i)f(x_i)$. By **(4)** the net $\{z_\alpha\}$ is contained in a weakly compact set. Hence there exists a subnet which converges weakly to a vector z in Y. But for each $y^* \in Y^*$ we have $x^*(z_\alpha) \to \int_G fd(x^*\mu)$. Thus $x^*(z) = \int_G fd(x^*\mu)$ which shows that $\int_G fd\mu = z$. Define $T : C_{rc}(S, X) \to Y$, $T(f) = \int_x fd\mu$. Then (i) T is u -continuous and (ii) weakly compact.

(i) For the continuity, let $q \in J$. Choose $p \in J$ such that $||\mu||_{p,q} < \infty$. If $x^* \in B_q^o \subseteq X^*$ and $||f||_p \leq 1$, we have

$$|x^*(T(f))| = \int_x fd(x^*\mu) \leq ||x^*\mu||_p \leq ||\mu||_{p,q}.$$

It follows that $||T||_{p,q} \leq ||\mu||_{p,q}$ and the continuity of T is established.

(ii) To prove the weak compactness consider an arbitrary bounded set A in $C_{rc}(S, X)$ and let H denote the convex circled hull of $\bigcup\{f(S) : f \in A\}$. Then H is bounded in X. Let

$$W = V_{m,H} = \{\Sigma\mu(G_i)x_i : \{G_i\} \text{ is a finite } B\text{-partition of } S, \ x_i \in H\} \subseteq Y$$

Clearly W is convex and circled. By hypothesis W is also weakly relatively compact in Y.. It follows that the polar W^o of W in Y^* is a $m(Y^*, Y)$ neighborhood of zero, $m(Y^*, Y)$ being the Mackey topology . We will show that $T^*(W^o) \subseteq A^o$. Let $x^* \in W^o$ and $f \in A$. If $G_1, ..., G_n$ is a $B(S)$-partition of S and $s_i \in G_i$, then $||x^*(\sum_1^n \mu(G_i)f(s_i))|| \leq 1$. This implies that

$$\left| x^*(\textstyle\int_S fd\mu) \right| \leq 1.$$
$$\text{Thus} \quad |\langle T^*x^*, f \rangle| = |\langle x^*, T(f) \rangle| \leq 1$$

which proves that $T^*x^* \in A^o$. Now the result follows from **Lemma 4.5.1**.
□

Note. By the preceding theorem, given a continuous weakly compact operator T from $C_{rc}(S, X)$ into Y, there exists $\mu \in M(B(S), L(X, Y))$ which represents T. Since the operator $\widehat{T} : C(\beta S, X) \to Y, \widehat{T}(\widehat{f}) = T(f)$, is also weakly compact and since the dual of $C(\beta S, X)$ (with the uniform topology) is $M_\tau(Bo(\beta S), X^*)$ we can find, using an argument analogous to that of **Lemma 4.5.2**, an $\widehat{\mu} \in M_\tau(Bo(\beta S), L(X, Y))$ representing $\widehat{T}$.

5.5   Notes and comments

We have seen in chapter 3 the important role played by the class of weakly compact operators on $C(S)$ in integral representation in Banach space settings [**BDS55**], [**DS58**]. This Chapter concerns two different representations for this class of operators in the context of locally convex spaces.

The first one, due to **D.R. Lewis** [**Lew70**], develops an appropriate integration process with respect to measures into a locally convex Hausdorff space X, and gives integral representation for a weakly compact operator $T : C(S) \longrightarrow X$, with respect to an $X-$valued measure in the case of a compact Hausdorff S. This extends the representation of **Bartle-Dunford-Schwartz** given in **Theorem 3.1.5** given for a Banach space X. If S is a locally compact Hausdorff space and $C_0(S)$ $[C_c(S)]$ is the Banach space under the sup. norm of scalar functions on S which vanish at infinity [have compact support], $C(S)_\beta$ denotes the space of bounded continuous scalar functions on S topologized by the the family of seminorms $p_\varphi(f) = \sup_{t \in S} |\varphi(t) f(t)|$, where $\varphi \in C_0(S)$. This is the strict topology introduced by **R. C. Buck** in [**Buc58**], in the more general setting of locally convex spaces. A weakly compact operator $A : C_0(S) \to X$ can be represented by integration with respect to $X-$valued measure since A can be extended to a weakly compact operator on the space of continuous functions over the one point compactification of S. Weakly compact operators on $C_c(S)$ have such a representationsince they can be extended to $C_0(S)$. The bounded sets of $C(S)_\beta$ are precisely the uniformly bounded sets and each element of $C(S)^*_\beta$ can be identified with a bounded regular Borel measure on S, [**Buc58**].

The second representation theorem is similar to that of Dinculeanu-Singer given in Theorem **3.2.1.** It is due to **A. Katsaras and D. B. Liu [KL 75]**, who considered an entirely different setting from that used by **Dinculeanu-Singer**. They have been able to get a representation for weakly compact operators, with respect to an operator valued measure for some special function spaces (see **Section 5.3** for details).

6

Integral Representations in General TVSs

This chapter presents a theory of integral representation whose settings goes beyond the locally convex TVS, namely to general TVS and especially to not locally convex TVS. As usual, the theory starts with the description of the integration process which will give the integral representation in an appropriate context. This theory has been developed by **A.H. Shuchat [Shu72]**. In his paper, Shuchat develops a Radon-type integral for vector valued functions with respect to operator valued measures,derived from that used by **Singer [Sing57]** and **Dinculeanu [Din67]** in Banach spaces.It includes generalization of Riesz representation theorem to this setting, using strong integrals. Basically the paper characterizes the continuous linear functionals linear functionals and operators on the space $C(S, X)$ of continuous vector functions with compact support defined on a locally compact space S (Riesz theorem).

6.1   Vector Measures and the Integration Process

Let S be a set, $A(S)$ an algebra of subsets of S, X and Y topological vector spaces (TVS) over either the real or complex field and μ an additive funciton on $A(S)$ with values in $L(X, Y)$, the space of continuous linear operators from X to Y. We wish to integrate X-valued functions on S and have the integral belong to Y.

A simple function: $f : S \to X$ is one with a finite number of values, each taken on a set in $A(S)$. For $A \in A(S)$ and $x \in X$ we denote the characteristic of A by χ_A and the function $t \to [\chi_A(t)]x$ from S to X by $\chi_A \otimes x$. The class $F_s(S, X)$ of simple functions is a vector space, each function $f = \Sigma_{i=1}^n \chi_{Ai} \otimes x_i$ is simple and each nonzero simple function has a unique representation in this form, where the x_i are distinct and the A_i are pairwise disjoint and cover S **[Din67, p.82]**. Such a family of sets, always finite, is called a $A(S)$-partition of S.

We define the integral of a simple function $f = \Sigma \chi_{Ai} \otimes x_i$ by

$$\int f d\mu = \sum \mu(A_i)x_i,$$

and, in particular,

$$\int \chi_A \otimes x \, d\mu = \mu(A)x.$$

The integral is independent of the representation chosen and is a linear transformation from $F_s(S, X)$ to Y [**Din67, p.108**]. Since each simple function is bounded, i.e., its range is a bounded set in X, topology of uniform convergence is a vector topology for $F_s(S, X)$. It follows immediately that integration is continuous on $F_s(S, X)$ if and only if for each neighborhood V of 0 in Y there is a neighborhood U of 0 in X such that if $x_1, ..., x_n \in U$ and $A_1, ..., A_n$ is an $A(S)$-partition of S, then

$$\Sigma \mu(A_i) x_i \in V. \qquad\qquad (\#BSV)$$

An additive set function from $A(S)$ to $L(X, Y)$ that satisfies (BSV) is called $L(X, Y)$-*valued measure* on $A(S)$ and $(\#BSV)$ is called the property of *bounded semi-variation*. Only finite additivity is a assumed. It is easy to verify that if X and Y are Banach spaces then μ is of bounded semivariation if and only if

$$\sup \|\Sigma \mu(A_i) x_i\|_Y < \infty,$$

where the supremum is taken over all $A(S)$-partitions of S and finite subsets of the closed unit ball in X. More generally, if X and Y are locally convex spaces then $(\#\mathbf{BSV})$ is equivalent to **Swong's property (V)** [**Swo64, p.275**]. With the algebraic operations defined in the usual way, the continuity property of $(\mathbf{BSV})$ shows that the $L(X, Y)$-valued measures form a vector space.

If Y is the scalar field then $\mu : A(S) \to X^*$ is of bounded semivariation if and only if, for some balanced neighborhood U of 0 in X,

$$
\begin{aligned}
Semivar_U \mu \;&=\; \sup \left\{ \left| \int f \, d\mu \right| : f \in F_s(S, X), \ f(S) \subseteq U \right\} \\
&=\; \sup \left| \sum \langle x_i, \mu(A_i) \rangle \right| = \sup \sum |\langle x_i, \mu(A_i) \rangle| \qquad (\#6.1)
\end{aligned}
$$

is finite, where the last two suprema are taken over all $A(S)$-partitions of S and elements of U and we use the inner product notation $\langle , \rangle$ for the pairing $\langle X, X^* \rangle$.

If X is a locally convex space whose topology is given by a family P of semi-norms, we say $\mu : A(S) \to X$ is of bounded variation if, for each $p \in P$,

$$\sup \Sigma p \left[\mu(A_i) \right] < \infty, \qquad\qquad (\#BV)$$

taken over all $A(S)$-partitions of S. Although we will only use this concept for locally convex spaces, the definition can be exteded to arbitrary spaces, using pseudometrics [**KN76**, p.52], and Proposition 6.2.1 below will remain valid.

For normed X, $\mu : A(S) \to X^*$ is of bounded semivariation if and only if it is of bounded variation, where the usual norm is used for X^* [**Tho68**]. For locally convex X, (BSV) implies (BV), where X^* has the strong topology [**Ula68**, p.412] and we present in §§ 6.2 that these results extended to certain nonlocally convex spaces.

Definition 6.1.1. (i) The **uniform topology** on the vector space $B(S,X)$ of all bounded functions from S to X is the usual vector topology.

(2) The closure $\overline{F_s}(S,X)$ of $F_s(S,X)$ in this space is called the **space of totally measurable functions**. Since the space of bounded functions are closed under uniform convergence, $\overline{F_s}(S,X)$ is the space of uniform limits of nets of simple functions and is complete if and only if X is complete.

(3) A **simple approximation** of $f \in \overline{F_s}(S,X)$ is a net of simple functions converging uniformly to f.

(4) A function from S to X is $A(S)$**-measurable** if the inverse image of each open set in X belongs to $A(S)$ and totally bounded if its range is a totally bounded set in X.

Proposition 6.1.2. (a) *Every totally measurable function is totally bounded.*

(b) *Every totally bounded $A(S)$-measurable function is totally measurable and the class of such functions is dense in $\overline{F_s}(S,X)$.*

Proof. (a) This assertion follows from [**KN76, p.70**].

(b) This assertion is true because every simple function is totally bounded and $A(S)$-measurable. $\square$

Remarks. (1) Now let f be totally bounded and $A(S)$-measurable. For each open neighborhood U of 0 in X, there are elements $t_1, ..., t_n$ of S such that the open sets $U_i = f(t_i) + U$ cover $f(S)$. Let $A_i = f^{-1}(U_i)$ and $B_1 = A_1$, $B_i = A_i - B_{i-1}$, $i = 2, ..., n$. Each $B_i \subseteq A_i$ and $\{B_i\}$ is a $A(S)$-partition of S. The simple function $s = \Sigma \chi_{B_i} \otimes f(t_i)$ has the property that $f(t) - s(t) \in U$ for each $t \in S$, so the result follows. Note that each such function f has a simple approximation $\{s_\alpha\}$ such that $s_\alpha(S) \subseteq f(S)$ for all α. This will be useful in the sequel.

(2) If Y is a complete Hausdorff space then the integral defined on $F_s(S,X)$ by an $L(X,Y)$-valued measure has a unique continuous linear extension to $\overline{F_s}(S,X)$. Thus, under these hypotheses, integraiton with respect to a vector measure is well defined for totally measurable functions.

(3) If T is an operator and μ an additive set function such that $Tf = \int f d\mu$ for all functions f in some class, then μ represents T on that class. The **Fichtenholz-Kantorovich-Hildebrandt theorem** in this setting is now easy to prove.

Theorem 6.1.3. [Shu72] If Y is a complete Hausdorff space then $L[\overline{F_s}(S,X), Y]$ is isomorphic to the space of $L(X,Y)$-valued measures on $A(S)$. Each operator in $L[\overline{F_s}(S,X), Y]$ is uniquely represented on $\overline{F_s}(S,X)$ by the corresponding measure.

Proof. If μ is an $L(X, Y)$-valued measure, let $\pi\mu \in L[\overline{F_s}(S, X), Y]$ be the integral determined by μ. By **(#6.1)**, the mapping π from the space of measures to $L[\overline{F_s}(S, X), Y]$ is one-one. If $T \in L[\overline{F_s}(S, X), Y]$ then

$$\mu(A)x = T[\chi_A \otimes x] \qquad (\#6.2)$$

defines an additive set function μ from $A(S)$ to $L(X, Y)$. Since T is continuous and **(#6.2)** implies that μ represents T on $F_s(S, X)$, μ is of bounded semivariation and thus a vector measure. By continuity, μ represents T on all of $\overline{F_s}(S, X)$, $\pi\mu = T$ and π is onto. Linearity is obvious, so the result follows. $\square$

Corollary. [Shu72] *The space of X^*-valued measures is isomorphic to $\overline{F_s}(S, X)^*$. Each functional in $\overline{F_s}(S, X)^*$ is uniquely represented by the measure to which it corresponds.* $\square$

Remark. We note that if no topological properties are assumed, i.e., if X and Y are vector spaces, then the proof of **Theorem 6.1.3** shows that the space of linear transformations from $F_s(S, X)$ to Y is isomorphic to the space of additive set functions on $A(S)$ whose values are linear trnsformations from X to Y.

Notations. Henceforth, S will denote a locally compact Hausdorff space and $\mathcal{B}o(S)$ the Borel class of S, i.e., the σ-algebra generated by the open sets.

Recall that: (1) A set function μ from $\mathcal{B}o(S)$ to a TVS X is **countably additive** if $\Sigma\mu(A_i)$ converges to $\mu(A)$ in X for each pairwise disjoint sequence (A_i) of Borel sets with union A.

(2) We say that μ is **regular** if, for every Borel set A and neighborhood U of 0 in X, there are Borel sets K and G such that

$$K \text{ is compact, } G \text{ is open,}$$
$$K \subseteq A \subseteq G \text{ and } \mu(B) \in U \text{ for every borel set } B \subseteq G - K. (\#6.3)$$

The next few results will be applied in §§ **6.2-6.3**. Let $\langle Z, Y \rangle$ be a pairing of vector spaces, Y have the weak topology $w(Y, Z)$, X be a TVS and μ be a set function from $\mathcal{B}o(S)$ to $L(X, Y)$. Each $\theta \in Z$ defines an X^*-valued set function μ_θ on $\mathcal{B}o(S)$ by

$$\langle x, \mu_\theta(A) \rangle = \langle \theta, \mu(A)x \rangle, \qquad (\#6.4)$$

where $A \in \mathcal{B}o(S)$ and $x \in X$. $L(X, Y)$ will have the simple topology of pointwise convergence.

We will be interested in two particular cases. **In § 6.3** we let Y be a TVS, $Z = Y^*$ and $Y = Y_\sigma^{**}$. (The bidual Y^{**} is the dual of Y_b^*, which is the dual space Y^* with the strong topology $b(Y^*, Y)$. Then Y_σ^{**} denotes Y^{**} with the weak topology $w(Y^{**}, Y^*)$.) Thus to $\mu : \mathcal{B}o(S) \to L(X, Y_\sigma^{**})$ there corresponds a family of set functions $\mu_\theta : \mathcal{B}o(S) \to X^*$, $\theta \in Y^*$. In § **6.2** we replace X by the scalar field $\mathbb{K}$, Z by X (with a given topology) and Y by X^*, the dual of X equipped with the weak topology $w(X^*, X)$. The space $L(\mathbb{K}, X^*_\sigma)$ is algebraiclaly isomorphic to $X^* = L(X, \mathbb{K})$ in the obvious way and so, by (**#6.4**), each $\mu : \mathcal{B}o(S) \to X^*$ determines a family of scalar-valued set functions μ_x, $x \in X$, defined by

$$\mu_x(A) = \langle x, u(A)A \rangle, \ A \in \mathcal{B}o(S). \tag{#6.5}$$

Proposition 6.1.4.. *In the above notation, μ is additive if and only if each μ_θ is additive and μ is countably additive (regular) for the simple topology on $L(X, Y)$ if and only if each μ_θ is countably additive (regular) for X^*_σ.*

Proof. The proofs of finite and countable additivity are straightforward and we omit them. The sets

$$N(x, V) = \{y \in L(X, Y) : yx \in V\}$$

form a local subbase in $L(X, Y)$ and x ranges over X and V ranges over a local subbase in Y. The sets

$$B(\theta, \varepsilon) = \{u \in Y : |u, \theta| < \varepsilon\}$$

form a local subbase in Y as θ ranges over Z and ε ranges over all positive numbers.

Then μ is regular if and only if, for each Borel set A and $U = N(x, V)$, where $V = B(\theta, \varepsilon)$, there are Borel sets K and G satisfying (**#6.3**). Also, $\mu(B) \in N(x, V)$ if and only if $|\langle x, \mu_\theta(B) \rangle| < \varepsilon$. Thus K and G satisfying (**#6.3**) for μ and $U = N(x, V)$ if and only if they satisfy (**#6.3**) for each μ_θ with $U = \{z \in X^* : |\langle x, z \rangle| < \varepsilon\}$. $\square$

Moreover, as the following result shows, μ is a measure if and only if each μ_θ is a measure.

Proposition 6.1.5. [Shu72] An additive set function μ is of bounded semivariation if and only if each μ_θ is of bounded semivariation.

Proof. We first note that if $s \in F_s(S, X)$ then

$$\left\langle \theta, \int s d\mu \right\rangle = \int s d\mu_\theta \tag{#6.6}$$

for all $\theta \in Z$, which follows directly from the case $s = \chi_A \otimes x$ and (**#6.4**). Now let $s_a \to 0$ in $F_s(S, X)$. Then $\int s_a d\mu \to 0$ in Y for $w(Y, Z)$ if and only if $\int s_a d\mu_\theta \to 0$ for each $\theta \in Z$. The result follows from the continuity characterization of (**BSV**).

We note for future reference that (**#6.6**) implies

$$\left\langle \theta, \int f d\mu \right\rangle = \int f d\mu_\theta \tag{#6.7}$$

for all $f \in \overline{F_s}(S, X)$, where $\int f d\mu$ belongs to the completion $\widehat{Y}$ of Y and the pairing $\left\langle Z, \widehat{Y} \right\rangle$ is derived in the obvious way from $\langle Z, Y \rangle$.

Since, for the simple topology, $L(\mathbb{K}, X_\sigma^*)$ and X_σ^* are isomorphic, we have the following corollary. Weak countable additivity and regularity refer to the topology of X_σ^*.

Corollary. [Shu72] *An X^*-valued set function μ is additive if and only if each scalar set function μ_x is additive and μ is weakly countably additive (regular) if and only if each μ_x is countably additive (regular).*

The case of **Proposition 6.1.5** is more delicate. We may regard X^* as being $L(X, \mathbb{K})$ or $L(\mathbb{K}, X_\sigma^*)$, but the notions of bounded semivariation differ because **(BSV)** depends on the individual topologies involved and not on a topology for the space of mappings. When considering **(BSV)** for X^*-valued set functions we will always regard X^* as being $L(X, \mathbb{K})$. With this in mind, we have the following results.

Proposition 6.1.6. [Shu72] *If an additive X^*-valued set function μ is of bounded semivariation then each scalar set function μ_x is of bounded variation. The converse false.*

Proof. It suffices to prove that integration with respect to each μ_x is continuous on $F_s(S, \mathbb{K})$. If s is a simple scalar function and $x \in X$ then $s \otimes x$ is a simple vector function and

$$\int s d\mu_x = \int s \otimes x d\mu. \qquad (\#6.8)$$

The assertion follows because if $s_a \to 0$ in $F_s(S, \mathbb{K})$ then $s_a \otimes x \to 0$ in $F_s(S, X)$. $\square$

However, if $X = L^1$ (for Lebesgue measure m on the Borel sets of $[0,1]$) and $\mu(A) = \chi_A \in L^\infty = X^*$, then μ is finitely additive and

$$\mu_x(A) = \int_A x dm,$$

so each μ_x is of bounded variation. But for any **Borel partition** of $[0,1]$ into n sets (with nonzero measure), $\Sigma \|\mu(A_i)\| = n$, so μ is not of strong bounded variation and, by the remarks following **(BSV)**, not of bounded semivariation.

Since each μ_x is countably additive, the preceding corollary implies that μ is weakly countably additive. However, it is clear that μ is not countably additive for the norm topology. This provides a counterexample to an assertion of **Singer [Sing57, p.303]**. The corollary to **Proposition 6.2.3** below shows that Singer's assertion becomes true in any TVS if μ is of strong bounded variation. A similar example can be used to show that the Orlicz-Pettis theorem fails for the weak* and norm topologies of the dual of a Banach space.

We note that if $\mu : \mathcal{B}o(S) \to L(X, Y)$ is additive then the mapping $\theta \to \mu_\theta$ defined by **(#6.4)** is a linear transformation from Z to the vector space of X^*-valued set functions on $\mathcal{B}o(S)$. Finally, it is easy to show that the properties of regularity, finite additivity and countable additivity are preserved in taking linear combinations of set functions. We denote by $M[L(X,Y)]$ the space of all regular countably additive $L(X,Y)$-valued Borel measures on S, using the simple topology on $L(X,Y)$ and the weak topology $w(Y,Z)$ on Y. In particular, $M(X^*)$ is the space of all weakly regular weakly countably additive X^*-valued Borel measures on S.

As for Banach spaces, problems concerning **(BV)** can be reduced to problems about positive measures. The proof of the next result is based on **[Din67, p. 33-36, 42, 318]** and the notation is that of **(BV)**.

Proposition 6.1.7. [Shu72] If an additive set function $\mu : \mathcal{B}o(S) \to X$ is of bounded variation then, for each $p \in P$,

$$\lambda_p(A) = \sup \sum p[\mu(A_i)],$$

taken over all $\mathcal{B}o(S)$-partitions of $A \in \mathcal{B}o(S)$, defines a finite positive and finitely additive set function on $\mathcal{B}o(S)$. λ_p is the least positive increasing superadditive function such that $p[\mu(A)] \leq \lambda_p(A)$. Also, μ is countably additive (regular) if and only if each λ_p is countably additive (regular).

The corollary below bears on the counterexample in **Proposition 6.1.6.**

Corollary. [Shu72] *If X and Y are TVS, $\mu : Bo(S) \to L(X,Y)$ is countably additive for the simple topology and of bounded variation for some topology t of uniform convergence on a family of sets that covers X, then μ is t-countably additive.*

Proof. If (A_i) is a pairwise disjoint sequence of Borel sets with union A, then since μ is of bounded variation for t, the partial sums of $\Sigma\mu(A_i)$ form a t-Cauchy sequence. But $\Sigma\mu(A_i)$ converges to $\mu(A)$ in the simple topology, and the t-neighborthoods of 0 are closed for the simple topology, so the result follows from the criterion in **[KN76, p.65]**.

6.2 $C_c(S,X)$ and its Dual Space

Let $C_c(S,X)$ denote the space of continuous functions from S to X with compact support. By **Proposition 6.1.2**, each such function is totally measurable with respect to the Borel sets of S, therefore integrable with respect to each vector-valued Borel measure on S, and we give $C_c(S,X)$ the uniform topology of $\overline{F_s}(S,X)$. When X is the scalar field we write $C(S)$. For each $a \in C(S)$ and $x \in X$ the function $t \to a(t)x$ from S to X, denoted by $a \otimes x$, belongs to $C_c(S,X)$ and the linear span of such functions is isomorphic to the algebraic tensor product $C(S) \otimes X$.

When X is locally convex $C(S) \otimes X$ is dense in $C_c(S,X)$, and in **[Shu72]** we show that this is true when X is a complete metrizable TVS with a basis (or satisfies a certain weaker condition) and also when X is an arbitrary TVS and S has finite covering dimension. It is not known if any space $C_c(S,X)$ fails to have this density property and for simplicity, we will assume throughout this article that it holds. Without this assumption, the representation theorems below may only hold for the closure of $C(S) \otimes X$.

The family of all convex balanced neighborhoods of 0 in X is a local base for a locally convex topology on X called the derived locally convex topology. We denote X with this topology by X_c. X_c has the strongest locally convex topology on X weaker than the original one, $(X_c)^* = X^*$, $X_c = X$ if and only if X is locally convex and X_c is a Hausdorff space if and only if X^* separates points of X [**KN76, p.109**]. As far as we know, the following result has appeared only for locally bounded spaces [**Vog67, p. 222**].

Proposition 6.2.1.. [**Shu72**] For every locally convex space Y, $L(X, Y) = L(X_c, Y)$.

Proof. Clearly, $L(X_c, Y) \subseteq L(X, Y)$. Conversely, if $T \in L(X, Y)$ and V is a convex neighborhood of 0 in Y, then the convex set $T^{-1}V$ contains a balanced neighborhood U of 0 in X and also its convex hull $\langle U \rangle$. Since $\langle U \rangle$ is balanced, T is also X_c-continuous.

Theorem 6.2.2. [**Shu72**] $C_c(S, X)^*$ is isomorphic to $M(X^*)$ and each functional in $C_c(S, X)^*$ is uniquely represented on $C_c(S, X)$ by the corresponding measure in $M(X^*)$.

Proof. By integration, each such X^*-valued measure defines a functional $T \in C_c(S, X)^*$ and the mapping $\pi : M(X^*) \to C_c(S, X)^*$ defined by $\pi\mu = T$ is linear, by the corollary to **Theorem 6.1.3**. For each $a \in C(S)$ and $x \in X$,

$$\int a \otimes x \, d\mu = \int a \, d\mu_x, \qquad (\#6.9)$$

obtained by approximating a by simple scalar functions and applying (**#6.8**). Thus if $\int f d\mu = 0$ for all $f \in C_c(S, X)$, then each scalar measure μ_x annihilates $C(S)$. Since μ is also weakly regular and weakly countably additive, the scalar **Riesz representation theorem** implies that each $\mu_x = 0$, so $\mu = 0$ and π is one-one.

It suffices to show that each $T \in C_c(S, X)^*$ is represented by an appropriate measure μ. For each nonzero $x \in X$, $C_x = \{a \otimes x : a \in C(S)\}$ is (algebraically isomorphic to $C(S)$. The restriction T_x of T to C_x determines a continuous linear functional on $C(S)$ and thus a unique regular countably additive scalar Borel measure μ_x on S such that

$$T[a \otimes x] = \int a \, d\mu_x , \qquad (\#6.10)$$

for all $a \in C(S)$. If $x = 0$, let $\mu_x = 0$. Define μ on the Borel sets of S by (**#6.5**). The values of μ are scalar functions on X. The linearity of T and the uniqueness of each μ_x imply that the mapping $x \to \mu_x$ is linear and therefore that μ is additive. In order to show that μ is X^*-valued, it suffices to prove that if $x_a \to 0$ in X then the corresponding scalar measures converge to 0 in variation, i.e., the mapping $x \to T_x$ from X to $C(S)^*$ is continuous , where $C(S)^*$ has the norm topology. Indeed, for each $\varepsilon > 0$, there is a balanced neighborhood U of 0 in X such that if $f(S) \subseteq U$ then

$|Tf| \leq \varepsilon$. For all $a \in C(S)$, $\|a\| \leq 1$, and large a, we have $[a \otimes x_a](S) \subseteq U$, so

$$\|Tx_a\| = \sup\left\{|T[a \otimes x_a]| : \|a\| \leq 1\right\} \leq \varepsilon.$$

In order to prove that μ is of bounded semivariation, let $\varepsilon > 0$, U be a balanced neighborhood of 0 in X as above, (A_i) be a Borel partition of S and (x_i) be a corresponding subset of U. Essentially following the procedure used by **Ulanov [Ula68, p. 420]**, we show that $|\Sigma \langle x_i, \mu(A_i) \rangle| \leq \varepsilon$ by, for any $\delta > 0$, combining the rgularity of each μ_{x_i} and of the space S and Urysohn's lemma to obtain functions a_i in $C(S)$ with norm one and pairwise disjoint supports such that

$$\left|\sum \langle x_i, \mu(A_i) \rangle\right| \leq \left|T\left[\sum a \otimes x_a\right]\right| + \delta. \qquad (\#6.11)$$

Thus μ is a weakly regular weakly countably additive X^*-valued Borel measure on S. The density property of $C_c(S, X)$, **(#6.9)** and **(#6.10)** imply that μ represents T on all of $C_c(S, X)$, so $T = \pi\mu$. $\square$

In certain cases the representing measures have stronger properties. In particular, if a representing measure is of strong bounded variation it is also strongly regular and strongly countably additive. Recall that the strong topology $b(X^*, X)$ of X_β^* is determined by the seminorms

$$p_M(\varphi) = \sup\left\{|\varphi(x)| : x \in M\right\}, \ \varphi \in X^*,$$

where M runs over all weakly bounded sets in X. .

Proposition 6.2.3. [Shu72] *Let $\mu : \mathcal{B}o(S) \to X^*$ be an additive set function of strong bounded variation.*

(a) For each weakly bounded set M in X the nonnegative finite set function

$$\lambda_M(A) = \sup \Sigma \, p_M[\mu(A_i)], \ A \in \mathcal{B}o(S),$$

taken over all partitions of A by Borel sets, is additive.

(b) μ is strongly countably additive (regular) if and only if μ is weakly countably additive (regular) if and only if each λ_M is countably additive (regular).

Proof. By **Proposition 6.1.7** and its corollary, the only thing left to prove is that weak regularity implies strong regularity, and the proof of this parallels that of Singer in Banach spaces **[Sing57, p.306]**.

We note that the strong properties mentioned here are equivalent to requiring that the measures μ_x satisfy the corresponding scalar properties uniformly over each weakly bounded set in X.

Recall that a measure is only assumed to be (finitely) additive and of bounded semivariation.

Corollary. [Shu72] *If every weakly bounded set in X is bounded, then every X^*-valued measure is of strong bounded variation and so the repre-*

senting measures in Theorem 6.2.2 are strongly regular and strongly countably additive. In particular, the result holds if

    (a) *the topology of X is weaker than the Mackey topology;*
    (b) *X is locally convex;*
    (c) *every linear functional on X is bounded;*
    (d) *every linear functional on X is continuous.*

Proof. if μ is an X^*-valued Borel measure on S then, by definition, μ is of bounded semivariation and so for some balanced neighborhood U of 0, Semivar $_U\mu < \infty$. Clearly, U can be replaced by any bounded set in X and so if each weakly bounded set M is bounded then

$$sup\Sigma\; p_M\left[\mu(A_i)\right] = \sup \Sigma \left|\langle x_i, \mu(A_i)\rangle\right| < \infty,$$

taken over all Borel partitions of S and elements of M, hence μ is of strong bounded variation.

Condition **(a)** is sufficient because then every Mackey bounded set is bounded. Condition **(b)** is, of course, a special case of **(a)**. $\square$

We will require the following numerical estimates in **§6.3**. Let μ be an X^*-valued Borel measure, U a closed balanced 0-neighbourhood in X for which Semivar $_U\mu < \infty$. For each $f \in \overline{F_s}(S, X)$,

$$N_U f = \inf \{b > 0 : f(S) \subseteq bU\}$$

is finite because f is bounded.

    Lemma 1. [Shu72] If $f \in \overline{F_s}(S, X)$ and $N_U f \neq 0$ then $f(S) \subseteq (N_U f)U$. Thus $N_U f \leq 1$ if and only if $f(S) \subseteq U$.

    Proof. For every $\varepsilon > 0$, $f(S) \subseteq (N_U f + \varepsilon)U$ because there exists $b > 0$ such that $N_U f < b < N_U f + \varepsilon$ and $f(S) \subseteq bU$. Thus, for each $t \in S$, x_ε and $f(S) \subseteq bU$. Thus, for each $t \in S$, $x_\varepsilon = (N_U f + \varepsilon)^{-1} f(t)$ belongs to U and, since U is closed, $(N_U f)^{-1} f(t)$ also belongs to U.

    Proposition 6.2.4. [Shu72] If $f \in \overline{F_s}(S, X)$ is $\mathcal{B}o(S)$-measurable, then $\left|\int f d\mu\right| \leq (N_U f)$ Semivar $_U\mu$.

    Proof. By **Proposition 6.1.2**, f has a simple approximation (s_a) such that $s_a(S) \subseteq f(S)$, hence $N_U s_a \leq N_U f$, for all a. Without loss, we may thus assume f is simple. If $N_U f \neq 0$, then **Lemma 1**,

$$\left|\int f d\mu\right| \leq (N_U f)\left|\int (N_U f)^{-1} f d\mu\right| \leq (N_U f) Semivar\; \mu.$$

If $N_U f = 0$ then, for all $b > 0$, $f(S) \subseteq bU$ and, by the above reasoning, $\left|\int f d\mu\right| \leq b$ Semivar $_U\mu$. So $\int f d\mu = 0$ and the result follows.

    Corollary. [Shu72] If $\mu \in M(X^*)$, then

$$Semivar\; _U\mu = \sup \left\{\left|\int f d\mu\right| : f \in C_c(S, X),\; N_U f \leq 1\right\}.$$

Proof. By the proposition, the right side above is no greater than the left.

The reverse inequality follows from the uniqueness of the representation in Theorem **6.2.2** and (#6.11).

6.3 Operators on $C(S, X))$

In this section we represent a continuous linear operator from $C(S, X)$ to a TVS Y whose dual separates points as an Y_n^{**}-valued integral by an $L(X, Y_n^{**})$-valued Borel measure on S. Y_n^{**}, the bidual Y^{**} equipped with the natural topology of uniform convergence on equicontinuous sets in Y^*, is a locally convex Hausdorff space whose topology is stronger than that of Y_σ^{**}. If Y is a normed space, then Y_n^{**} has the usual norm topology of Y^{**} [**Sch71, p.143**]. By Y_n^{**}-valued integral, we mean that the integral is defined as a limit in the space Y_n^{**}. Thus this is a strong rather than weak representation. By placing certain additional restrictions on the measure, the representation turns out to be unique. However, because the methods used depend on the pairing $\langle Y, Y^* \rangle$ and the locally convex topologies it determines, the representation does not distinguish the mappings in $L[C(S, X), Y]$ from those in the potentially larger space $L[C(S, X), Y_c]$. In general, this is a nontrivial distinction, as the following example shows.

Let X be nonlocally convex and S compact, so that the function 1 (identically equal to one) belongs to $C(S)$. The evaluation map T at any fixed point in S belongs to $L[C(S, X), Y_c]$. However, if $x_a \to 0$ in X_c but not in X then $1 \otimes x_a \to 0$ in $C(S, X_c)$ but its image under T diverges in X, so T does not belongs to $L[C(S, X), Y_c]$.

If Y^* separates points of Y then Y may be viewed as a subspace of Y^{**} and when Y is locally induced by Y_n^{**} is the original one of Y.

Proposition 6.3.1. [**Shu72**] If Y is a TVS whose dual separates points, then the topology induced on Y by Y_n^{**} is that of Y_c.

Proof. Since Y_c is locally convex, it suffices to show that the topologies of Y and Y_c determine the same equicontinuous sets in Y^*. Since Y_c is weaker than Y, each Y_c-equicontinuous set is Y-equicontinuous. Conversely, if $\theta \in Y^*$ maps a balanced neighborhood of 0 in Y into a ball B in the scalar field then θ also maps its convex hull, a neighborhood of 0 in Y_c, into B. The result follows.

A family S of X^*-valued measures is of **uniformly bounded semivariation** if there is a neighborhood U of 0 in X such that

$$\sup\{\text{Semivar }_U \lambda : \lambda \in S\} < \infty.$$

Without loss, we may take U closed and balanced.

Proposition 6.3.2. [**Shu72**] *A subset of $M(X^*)$ is equicontinuous as a family of functions on $C(S, X)$ if and only if it is of uniformly bounded semivariation.*

Proof. By the corollary to **Proposition 6.2.4**, Semivar$_U\lambda$ can be computed using continuous functions. The necessity of the condition follows easily, using **Lemma 1**.

For sufficiency, if Semivar$_U\lambda$ is bounded on S, then for every $\varepsilon > 0$, $\lambda \in S$ and $f \in C(S, X)$ such that $f(S) \subseteq \varepsilon U$, $\left|\int f d\lambda\right| \leq \varepsilon.$Semivar$_U\lambda$, so S is equicontinuous. $\square$

Theorem 6.3.3. [**Shu72**] *There is an isomorphism of* $L\left[C(S, X), Y\right]$ *into the subspace* $M_n\left[L(X, Y_\sigma^{**})\right]$ *for which*

(**#6.12**) *the mapping* $\theta \to \mu_\theta$ *from* Y^* *to* $M(X^*)$ *is weakly continuous and takes equicontinuous sets to families of uniformly bounded semivariation.*

Each such mesure is also an $L(X, Y_n^{**})$*-valued measure and each operator in* $L[C(S, X), Y]$ *is uniquely represented as an* Y_n^{**}*-valued integral by the corresponding measure.*

Remark. (**#6.12**) referes to the pairings $\langle Y^*, Y \rangle$ and $\langle M(X^*), C(S, X) \rangle$. If we replace Y by Y_c (equivalently, if Y is locally convex), the isomorphism is onto. It is easy to see that every $L(X, Y_n^{**})$-valued measure is an $L(X, Y_\sigma^{**})$-valued measure. Since $\widehat{Y}_n^{**}$ (the completion of Y_n^{**}) can be embedded in $\widehat{Y}_\sigma^{**}$ with a resultant weakening of its topology [**KN76**, p.65], the $\widehat{Y}_n^{**}$- and $\widehat{Y}_\sigma^{**}$-valued integrals determined by such a measure coincide on $\overline{F_s}(S, X)$. Part of the assertion of the theorem is that, for $f \in C(S, X)$, $\int f d\mu$ belongs to Y and converges in Y_n^{**} (not just $\widehat{Y}_n^{**}$).

Before proving the theorem, we prove the following lemma.

Lemma 2. [**Shu72**] Each measure in $M_n\left[L(X, Y_\sigma^{**})\right]$ is an $L(X, Y_n^{**})$-valued measure.

Proof. In order to see that $\mu \in M_n\left[L(X, Y_\sigma^{**})\right]$ is $L(X, Y_n^{**})$-valued, let A be a Borel set, $x_a \to 0$ in X and $M \subseteq Y^*$ be equicontinuous. By (#6.12), there is a closed balanced neighborhood U of 0 in X such that Semivar$_U\mu_\theta$ is bounded for $\theta \in M$. For each such θ,

$$|\langle\theta, \mu(A)x_a\rangle| \quad = \quad |\langle x_a, \mu_\theta(A)\rangle| \leq N_U[\chi_A \otimes x_a]\,\text{Semivar}_U\mu_\theta,$$

and $N_U[\chi_A \otimes x_a] \quad \to \quad 0$, so $\mu(A)x_a \to 0$ uniformly on M and thus in Y_n^{**}.

Finally, μ is an $L(X, Y_n^{**})$-valued measure, i.e., is of bounded semivariation for Y_n^{**}. It suffices to show that if $s_a \to 0$ in $F_s(S, X)$, then $\langle\theta, \int s_a d\mu\rangle \to 0$ uniformly on every equicontinuous set $M \subseteq Y^*$. But if U is as above,

$$\left|\left\langle\theta, \int s_a d\mu\right\rangle\right| = \left|\int s_a d\mu_\theta\right| \leq N_U s_a\text{Semivar}_U\mu_\theta,$$

so the assertion follows.

The measures in Lemma 2 may not be regular or countably additive for a nice topology on $L(X, Y_n^{**})$, as an example of Dinculeanu [**Din67, p.401**] in Banach spaces shows. In this connection, see Proposition **6.3.4** below.

Proof of Theorem 6.3.3. [Shu72] Let $T \in L[C(S,X), Y]$. We will define a measure $\mu : \mathcal{B}o(S) \to L(X, Y_n^{**})$ that satisfies (#6.12) and represents T as follows. We first note that $T^{**} : C(S,X)_\sigma^{**} \to Y_\sigma^{**}$ is well defined and continuous, then prove that $T^{**}f$ makes sense for $f \in F_s(S,X)$ jand finally define

$$\mu(A)x = T^{**}[\chi_A \otimes x], \qquad (\#6.13)$$

for $A \in \mathcal{B}o(S)$, $x \in X$. This defines $\mu(A)$ as a continuous linear mapping from X to Y_σ^{**}. Next, $\mu \in M[L(X, Y_\sigma^{**})]$ satisfies (**#6.12**) and represents T as an Y_σ^{**}-valued integral. Finally, Lemma 2 and the remark imply $Tf = \int f d\mu$ as an Y_n^{**}-valued integral for all $f \in C(S,X)$.

We write $M(X^*)$ for $C(S,X)^*$ and $M(X^*)^*$ for $C(S,X)^{**}$. By [**Sch71, pp.129, 158**], the existence of the adjoint map $T^* : Y^* \to M(X^*)$ does not depend on convexity, T^* is both weakly and strongly continuous and $T^{**} : M(X^*)^* \to Y^{**}$ exists and is weakly continuous.

Each $f \in \overline{F_s}(S,X)$ determines, by integration, a linear functional (again denoted by f) on $M(X^*)$ and this mapping from $\overline{F_s}(S,X)$ to the algebraic dual of $M(X^*)$ is linear, although perhaps not one-one. If $f \in C(S,X)$ then f is of course strongly continuous, i.e., an element of $M(X^*)^*$, and we will show that this is also true of simple functions.

It suffices to show that each $\chi_A \otimes x$ is strongly continuous and to this end, let $\lambda_\alpha \to 0$ in $M(X^*)_\beta^*$ and $\varepsilon > 0$. Since each $(\lambda_\alpha)_x$ is a regular scalar-valued Borel measure, there is a net (φ_α) in $C(S)$ such that $\|\varphi_\alpha\| \leq 1$ and (see (**#6.9**))

$$\left| \int \chi_A \otimes x.d\lambda_\alpha \right| \leq \left| \int \varphi_\alpha \otimes x d\lambda_\alpha \right| + \varepsilon. \qquad (\#6.14)$$

By the definition of convergence in $M(X^*)_\beta^*$, if the set $\{\varphi_\alpha \otimes x\}$ is weakly bounded in $C(S,X)$, then the right side of (**#6.14**) can be made arbitrary close to ε by taking a arbitrarily large, and this willl prove that $\chi_A \otimes x$ is strongly continuous. For this, let $\lambda \in M(X^*)$ and U be a closed balanced neighborhood of 0 in X such that $\text{Semivar}_U \lambda$ is finite. For each a,

$$\left| \int \varphi_\alpha \otimes d\lambda_\alpha \right| \leq N_U(\varphi_\alpha \otimes x)Semivar_U \lambda \leq N_U(1 \otimes x)Semivar_U \lambda,$$

so $\{\varphi_\alpha \otimes x\}$ is weakly bounded.

It now follows that for $f \in F_s(S,X)$, both $T^{**}f$ and (**#6.13**) make sense. Clearly, each $\mu(A)$ is linear, $\mu(A)$ is linear, μ is additive and $T^{**}f = \int f d\mu$ for each $f \in F_s(S,X)$.

Each $\mu(A) \in L(X, Y_\sigma^{**})$ because if $x_a \to 0$ in X, then $\chi_A \otimes x_a \to 0$ uniformly and thus in $M(X^*)_\sigma'$ and, since T^{**} is weakly continuous, $\mu(A)x_a \to 0$ in Y_σ^{**}.

For the next step, note that, for each Borel set A, $\theta \in Y^*$ and $x \in X$,

$$\begin{aligned}
\langle x, \mu_\theta(A) \rangle &= \langle \theta, T^{**}[\chi_A \otimes x] \rangle = \langle T^*\theta, \chi_A \otimes x \rangle \\
&= \int \chi_A \otimes x \, d\lambda = \langle x, \lambda(A) \rangle,
\end{aligned}$$

where $\lambda = T^*\theta \in M(X^*)$. Thus the mapping $\theta \to \mu_\theta$ is just T^*, each $\mu_\theta \in M(X^*)$ and, by **Propositions 6.1.4** and **6.1.5**, $\mu \in M[L(X, Y_\sigma^{**})]$. T^* is weakly continuous and [**Sch71, p.130**] preserves equicontinuity so, by **Proposition 6.3.2** and the proof of **Proposition 6.3.1**, μ satisfies (**#6.12**). In order to prove that μ represents T as an Y_σ^{**}-valued integral, first note that, by (**#6.13**), μ represents T^{**} on $F_s(S, X)$. If (s_a) is a simple approximation of $f \in C(S, X)$, then $s_a \to f$ in $M(X^*)_\sigma^*$. Since T^{**} is weakly continuous, $\int f d\mu$ belongs to Y^{**} and equals $T^{**}f$. Since the restriction of T^{**} to (the image of) $C(S, X)$ is T, $Tf = \int f d\mu$.

By **Lemma 2** and the remark, this part of the proof is complete.

We next show that every measure $\mu \in M_n[L(X, Y_\sigma^{**})]$ determines, by integration, an operator in $L(C(S, X), Y_c]$. By **Lemma 2** and the remark, T belongs to $L(C(S, X), \widehat{Y}_n^{**}]$, so by **Proposition 6.3.1**, it suffices to show that, for each $f \in C(S, X)$, $\int f d\mu$ belongs not only to $\widehat{Y}_n^{**}$ but actually to Y. It thus suffices to show that $\int f d\mu$ is continuous on Y_σ^* and to this end, let $\theta_a \to 0$ in Y_σ^*. Since $\theta \to \mu_\theta$ is weakly continuous, (**#6.7**) implies that $\int f d\mu\theta_a \to 0$ and so $T \in L(C(S, X), Y_c]$. Moreover, the measure obtained from T by (**#6.13**) is precisely μ, because the representation is unique. Indeed, if $\mu \in M[L, Y_\sigma^{**})]$ satisfies (**#6.12**) and $\int f d\mu = 0$ for all $f \in C(S, X)$, then, for all $\theta \in Y^*$, $\int f d\mu = 0$. By the uniqueness in **Theorem 6.2.2**, each $\mu_\theta = 0$ and so $\mu = 0$.

Thus the range of the mapping $\pi : L[C(S, X), Y] \to M[L(X, Y_\sigma^{**})]$ defined by (**#6.13**) is contianed in the subspace $M_n[L(X, Y_\sigma^{**})]$ satisfying (**#6.12**), and if we define π on the larger space $L[C(S, X), Y_c]$, then π is onto. Obviously, π is onto. Obviously, π is one-one and, since the taking of adjoints is linear, π is linear. So π is an isomorphism and the proof of the theorem is complete.

Corollary 1. [**Shu72**] If Y_c is semireflexive then the representing measures in **Theorem 6.3.3** are $L(X, Y_c)$-valued measures and the corresponding integrals converge in Y_c.

Corollary 2. [**Shu72**] In addition to the hypotheses of **Theorem 6.3.3**, let X^* separate points of X and Y_c be quasicomplete. Then the operators in $L[C(S, X), Y_c]$ that are weakly compact in $L[C(S, X)_c, Y_c]$ correspond to the $L(X, Y_c)$-valued representing measures for which the mapping $\theta \to \mu_\theta$ is continuous for Y_σ^* and $M(X^*)$ with the weak topology $w(M(X^*), M(X^*)^*)$.

Proof. Note that **Proposition 6.2.1** implies $L[C(S, X), Y_c] \subseteq L[C(S, X)_c, Y_c]$. This corollary is a standard application of [**Bou, p.89**] as in **Swong** [**Swo64, p.285**], once we observe that if X^* separates points of X then $M(X^*)$ separates points of $C(S, X)$, since $C(S, X)_c$ must be a Hausdorff space. Indeed, since X^* separates points of X, X_c is a Hausdorff space and so is the locally convex space $C(S, X_c)$, whose dual $M(X'_c)$ thus also separates points. Also, $C(S, X) \subseteq C(S, X_c)$, $M(X'_c) \subseteq M(X^*)$ and if $f \in C(S, X)$ and $\mu \in M(X'_c)$ then $\int f d\mu$ has the same meaning if we regard μ as being in $M(X^*)$. Thus $M(X^*)$ separates points of $C(S, X)$ and the proof is complete.

Proposition 6.3.4. [Shu72] If a representing measure $\mu \in M_n[L(X, Y_\sigma^{**})]$ is of bounded variation for the topology of bounded convergence on $L(X, Y_n^{**})$, then μ is countably additive and regular for that topology.

Proof. The topology of $L(X, Y_n^{**})$ is given by the seminorms

$$p_{MB}(\mu) = \sup\{p_M(ux) : x \in B\},$$

where $B \subseteq X$ is bounded, $M \subseteq Y^*$ is equicontinuous and

$$p_M(\phi) = \sup\{|\langle \theta, \phi \rangle| : \theta \in Y^*\}$$

is the corresponding seminorm for Y_n^{**}. Let λ_{MB} be the corresponding positive measure.

For countable additivity, we note that if (A_i) is a sequence of pairwise disjoint Borel sets with union A, then the partial sums of $\Sigma\mu(A_i)$ form a Cauchy sequence in $L(X, Y_n^{**})$ for the topology of bounded convergence. Moreover, it is easy to see that the neighborhoods of 0 determined by the p_{MB} are closed for the restriction to $L(X, Y_n^{**})$ of the simple topology of $L(X, Y_\sigma^{**})$. Since $\Sigma\mu(A_i)$ converges to $\mu(A)$ in the latter topology, the result follows as in the corollary to **Proposition 6.1.7.**

For regularity, fix some M and B, let A be a Borel set and $\varepsilon > 0$. Let $A_1, ..., A_n$ be a Borel partition of A such that

$$\lambda_{MB}(A) - \sum p_{MB}[\mu(A_i)] < \varepsilon,$$

$x_1, ..., x_n \in B$ such that each

$$p_{MB}[\mu(A_i)] - p_M[\mu(A_i)x_i] < \varepsilon/n,$$

and $\theta_1, ..., \theta_n$ such that each

$$p_M[\mu(A_i)x_i] - |\langle \theta_i, \mu(A_i)x_i \rangle| < \varepsilon/n.$$

Suppress the indices on the vectors and recall that each $(\mu_\theta)_x$, defined by (#6.4) and (#6.5), is a regular scalar measure, so that there are compact sets $K_i \subseteq A_i$ such that

$$|(\mu_\theta)_x(A_i)| - |(\mu_\theta)_x(K_i)| < \varepsilon/n.$$

Let $K = \bigcup K_i$ and paste these relations together (with certain additional obvious ones as in **[Sing57, p. 306]**) to obtain

$$\lambda_{MB}(A) - \lambda_{MB}(K) < 4\varepsilon.$$

The existence of a suitable open set G containing A is obtained by duality.
$\square$

6.4 Notes and comments

In recent years, increasing attention has been paid to topological vector spaces that are not locally convex and to vector-valued measures and integrals.The aim of this chapter is to present an integral representation theory for operators on function spaces in the setting of general topological vector spaces. We adopted the point of view of **A H Shuchat**. In his paper [**Shu72**], Shuchat develops in § **6.1** a Radon-type integral for vector valued functions with respect to operator valued measures,derived from that used by Singer [**Sing57**] and Dinculeanu [**Din67**] in Banach spaces.This often seems to lend itself to "softer" proofs than the Riemann-Stieltjes-type integal used by **Swong** [**Swo64**] and Goodrich [**Goo70**] in locally convex spaces. Since the spaces need not be locally convex, the basic integration process avoids the use of the Hahn-Banach theorem as in [**Din67**], [**Sin57**], for example.

As usual in any integral representation theory, the paper characterizes, the continuous linear functionals linear functionals and operators on the space $C(S, X)$ of continuous vector functions with compact support defined on a locally compact space S (Riesz theorem). Strong integral representations are obtained, as opposed to the weak representations of **Foias and Singer** [**FS60**] in Banach spaces and **Ulanov** [**Ula68**] and **Swong** [**Swo64**] in locally convex spaces.

We refer to Sections **A.3-A.4** and also the books [**KN76**] and [**Sch71**] for general facts about topological vector spaces.

Appendix A

Functional Analysis and Measure Theory

In this Appendix, we present preliminaries on topological spaces, Banach spaces, topological vector spaces, duality theory, spaces of continuous vector-valued functions and measure theory for reference purpose and convenience of the readers.

A.1 Topology

In this section we present some basic definitions and results on topological spaces.

Definition: Let (S, τ) be a topological space and $A \subseteq S$.

(1) A point x in A is called an **interior point** of A if there exists an open set U in S such that $x \in U \subseteq A$.

(2) A point $x \in A$ is called an **isolated point** of A if it has a neighborhood which does not contain any other point of A.

(3) A point $x \in S$ is called a **limit point** of A if each neighborhood of x contains a point of A other than x. The set of all limit points of A is denoted by A_d.

Note that, for any $A \subseteq S$, its closure $\overline{A} = A \bigcup A_d$; so A is closed iff every limit point of A is in A. Further, every $x \in \overline{A}$ is either an isolated point of A or a limit point of A.

Definition: (1) A relation $\geq$ on a set D is called a **direction** *(or a directed relation)* if:

(i) $\alpha \geq \alpha$, for all $\alpha \in D$ (reflexive),

(ii) if $\alpha \geq \beta$, $\beta \geq \gamma$, then $\alpha \geq \gamma$ (transitive);

(iii) if $\alpha, \beta \in D$, there exists a $\gamma \in D$ such that $\gamma \geq \alpha$, $\gamma \geq \beta$ (directive).

In this case, $(D, \geq)$ is called a *directed set*.

(2) A **net** in a set S is a mapping f of a directed set D into S. We denote this net by $\{f(\alpha) : \alpha \in D\}$ or $\{x_\alpha : \alpha \in D\}$, where $x_\alpha = f(\alpha) \in S$. If B is a subset of a directed set $(D, \geq)$, with the induced ordering is directed, then $(B, \geq)$ is called a **directed subset** of $(D, \geq)$. A *subnet* $\{x_\beta : \beta \in B\}$ in S is a directed subset of a net $\{x_\alpha : \alpha \in D\}$.

(3) A net $\{x_\alpha : \alpha \in D\}$ in a topological space (S, τ) is said to be *convergent* to $x \in S$ if, given any neighborhood U of x, there exists $\alpha_0 \in D$ such that

$$x_\alpha \in U \quad \text{for all } \alpha \geq \alpha_0,$$

(i.e., $x_\alpha \in U$ **eventually**). In this case, x is called the **limit of the net** and we write $x_\alpha \longrightarrow x$.

Theorem A.1.2. *Let S be a topological space and $A \subseteq S$. Then:*

(a) For any $x \in S, x \in \overline{A}$ iff there exists a net $\{x_\alpha\} \subseteq A$ such that $x_\alpha \longrightarrow x$.

(b) A is closed iff, for any net $\{x_\alpha\} \subseteq A$ with $x_\alpha \to x \in S$, we have $x \in A$.

Definition: Let S and T be topological spaces.

(1) A function $f : S \to T$ is said to be **continuous at** $x_0 \in S$ if, for each neighborhood V_0 of $f(x_0)$ in T, there exists a neighborhood U_0 of x_0 in S such that $f(U_0) \subseteq V_0$. f is said to be **continuous on** S if it is continuous at each $x \in S$.

(2) A function $f : S \to T$ is called a **homeomorphism** if f is continuous, one-one, onto and $f^{-1} : T \to S$ is continuous.

It is often convenient to use the following equivalent conditions for continuity.

Theorem A.1.3. *Let S and T be topological spaces and $f : S \to T$ a function.*

(a) The following conditions are equivalent:

(i) $f : S \to T$ is continuous on S.

(ii) For each open set $V \subseteq T, f^{-1}(V)$ is open in S.

(iii) For each closed set $W \subseteq T, f^{-1}(W)$ is closed in S.

(iv) For any net $\{x_\alpha\} \subseteq S$ with $x_\alpha \to x$ in $S, f(x_\alpha) \to f(x)$ in T.

(b) If S and T are metric spaces, then $f : S \to T$ is continuous on S iff, for any sequence $\{x_n\} \subseteq S$ with $x_n \to x$ in $S, f(x_n) \to f(x)$ in T.

(c) If τ_1 and τ_2 are two topologies on a set S, then the identity map $i : (S, \tau_1) \to (S, \tau_2)$ is continuous iff $\tau_2 \subseteq \tau_1$.

Definition. A mapping $\varphi : (S, d) \longrightarrow (T, \rho)$ is called an ***isometry*** of (S, d) into (T, ρ) if

$$\rho(\varphi(x_1), \varphi(x_2)) = d(x_1, x_2) \text{ for all } x_1, x_2 \in S.$$

(i.e. φ preserves distances). In this case, S and $\varphi(S)$ are said to be ***isometric***.

Remarks. (1) Clearly, every isometry is one-one, continuous and an open mapping.

(2) Hence an isometry $\varphi : (S, d) \to (T, \rho)$ is a homeomorphism $\Longleftrightarrow \varphi$ is onto.

(3) However, a homeomorphism $\varphi : (S, d) \to (T, \rho)$ need not be an isometry.

Definition: If S, T, Z are topological spaces, then a function $f : S \times T \to Z$ is called

(i) **separately continuous** on $S \times T$ if, given any $(x, y) \in S \times T$ and any neighborhood G of $f(x, y)$ in Z, there exist neighborhoods H of x in S and J of y in T such that $f(H \times \{y\}) \subseteq G$ and $f(\{x\} \times J) \subseteq G$.

(ii) *jointly continuous* on $S \times T$ if, given any $(x, y) \in S \times T$ and any neighborhood G of $f(x, y)$ in Z, there exist neighborhoods H of x in S and J of y in T such that $f(H \times J) \subseteq G$.

Clearly, jointly continuity $\Rightarrow$ separately continuity; the converse need not hold.

Definition: A topological space (S, τ) is called:

(1) *Hausdorff* if, for any two points $x \neq y$ in S, there exist two open sets $U, V \subseteq S$ such that $x \in U$, $y \in V$ and $U \cap V = \varnothing$;

(2) **normal** if for any two closed sets $A, B \subseteq S$ with $A \cap B = \varnothing$, there exist two open sets $U, V \subseteq S$ such that $A \subseteq U, B \subseteq V$ and $U \cap V = \varnothing$;

(3) *completely regular* if, for any closed set $A \subseteq S$ and $x \in S$ with $x \notin A$, there exists a continuous function $f : S \to [0, 1]$ such that $f(x) = 0$ and $f(y) = 1$ for all $y \in A$.

Definition: Let S be a topological space, and let $A \subseteq S$.

(1) A collection $\mathcal{U} = \{G_\alpha : \alpha \in I\}$ of open subsets of S is said to be an **open cover** of A if $A \subseteq \bigcup_{\alpha \in I} G_\alpha$. If $\mathcal{U}$ *and* $\mathcal{V}$ are two covers of S, then $\mathcal{V}$ is said to be a **refinement** of $\mathcal{U}$ if each $V \in \mathcal{V}$ is contained in some $U \in \mathcal{U}$.

(2) A is called **compact** if every open cover of A has a finite subcover.

(3) A is called **sequentially compact** if every sequence in A has a convergent subsequence with limit in A.

(5) A is called **relatively compact** if its closure $\overline{A}$ is compact.

(5) If A is a subset of a pseudometric space (S, d), then A is called **precompact** (or *totally bounded*) if, for each $\varepsilon > 0$, there exists a finite subset $D = \{x_1, ..., x_n\}$ of S such that

$$A \subseteq \bigcup_{i=1}^{n} B(x_i, \varepsilon).$$

Definition: A topological space (S, τ) is called:

(1) **locally compact** if each $x \in S$ has a relative compact neighborhood;

(2) *pseudocompact* if every continuous real-valued function on S is bounded;

(3) σ**-compact** if it can be written as a countable union of compact sets.

(4) *a k-space* if, for any $U \subseteq S$, U is closed in S whenever $U \cap K$ is closed in K for each compact $K \subseteq S$;

(5) *a $k_\mathbb{R}$-space* if *a* function $f : S \to \mathbb{R}$ is continuous on S whenever $f \mid K$ is continuous for every compact subset K of S;

(6) **separable** if it has a countable dense subset.

We now summarize some properties of the above spaces in the following theorem.

Theorem A.1.4. *(a) A topological space S is Hausdorff iff every convergent net in S has a unique limit.*

(b) A closed subset A of a compact space S is compact.

(c) A compact subset A of a Hausdorff space S is closed.

(d) Every compact Hausdorff space is normal.

(e) A subset A of a topological space S is compact iff every net in A has a convergent subnet with limit in A.

(f) Every compact space and every sequentially compact space is pseudo-compact.

(g) A pseudometric space is compact iff it is sequentially compact iff it is countably compact iff it is precompact and complete.

(h) Every locally compact space and every first countable (in particular, metric) space is a k-space.

(i) The continuous image of a compact set is compact.

(j) The continuous image of a separable space is separable.

(k) If S is a k-space and T any topological space, then a function $f : S \to T$ is continuous on S iff $f \mid K$ is continuous for every compact subset K of S; hence every k-space is a $k_{\mathbb{R}}$-space.

(l) Every compact metric space is separable.

Note. Clearly, a finite subset A of a topological space S is always compact; but need not be closed. So it need not be relatively compact (e.g. if S is neither T_1 nor regular).

Example. Let $S = \mathbb{N}$ with the topology

$$\tau = \{\mathbb{N}\}\bigcup\{A_n : n \in \mathbb{N}\},$$

where $A_n = \{1, 2, ...n - 1\}$ for every $n \in \mathbb{N}$. This is T_0 but not T_1 (hence also not Hausdorff). In this case, the only closed compact subsets of S and $\varnothing$, so the only relatively compact set is $\varnothing$.

Theorem A.1.5. *Let S be any topological space and T a Hausdorff topological space, and let $f, g : S \to T$ be continuous. Then:*

(a) The set $A = \{x \in S : f(x) = g(x)\}$ is closed in S.

(b) If $D \subseteq S$ and $f = g$ on D, then $f = g$ on $\overline{D}$. In particular, if D is dense in S and $f = g$ on D, then $f = g$ on S.

Notation. For any function $\varphi : S \to \mathbb{R}$, $A \subseteq S$, and $t \in \mathbb{R}$, we shall often write $\varphi(A) = t$ to mean that $\varphi(A) = \{t\}$, that is, $\varphi(x) = t$ for all $x \in A$.

Theorem A.1.6. (Urysohn Lemma) *Let S be a normal space, and let A and B be closed subsets of S with $A \cap B = \varnothing$. Then there exists a continuous function $\varphi : S \to [0, 1]$ such that $\varphi(A) = 0$ and $\varphi(B) = 1$.*

Theorem A.1.7. (a) *Let S be a completely regular space, K a compact subset of S, and B a closed of S with $K \cap B = \varnothing$. Then there exists a continuous function $\varphi : S \to \mathbb{R}$ such that $\varphi(K) = 0$ and $\varphi(B) = 1$.*

(b) *Let S be a locally compact Hausdorff space, K a compact subset of S, and U a neighborhood of K. Then there exists a continuous function $\varphi : S \to [0, 1]$ such that $\varphi(K) = 1$ and the support of φ is compact and contained in U. (The support of φ is the closure of the set $\{x \in S : \varphi(x) \neq 0\}$).*

In particular, every normal T_1-space and every locally compact Hausdorff space is completely regular. Further, every 0-dimensional T_1-space is completely regular.

Definition: A topological space S is called a **Baire space** if whenever $S = \bigcup_{n=1}^{\infty} A_n$ with each A_n closed, then, for at least one m, $int(A_m) \neq \varnothing$.

Theorem A.1.8. (Baire) *(a) A topological space S is a Baire space iff, for any countable family $\{U_n\}$ of open dense subsets in S, $\cap_{n=1}^{\infty} U_n \neq \varnothing$.*

(b) Every locally compact Hausdorff space is a Baire space.

(c) Every complete metric space is a Baire space.

Theorem A.1.9. (Tietze extension Theorem) *Let S be a normal space, and let A be a closed subset of S. Then, for any continuous function $f : A \to [0,1]$ (or $\mathbb{R}$), there exists a continuous function $g : S \to [0,1]$ (or $\mathbb{R}$) such that $g = f$ on A.* (Here g is called a *continuous extension* of f from A to S).

Definition: Let $\{(S_\alpha, \tau_\alpha) : \alpha \in I\}$ be an indexed family of topological spaces. The **product topology** on the Cartesian product $\underset{\alpha \in I}{\Pi} S_\alpha$ is defined as the topology τ_Π obtained by taking as its base the collection of all sets of the form $\underset{\alpha \in I}{\Pi} U_\alpha$, where each U_α is τ_α-open in S_α and $U_\alpha = S_\alpha$ except for a finite number of indices α's.

Tychonoff theorem. If $\{S_\alpha : \alpha \in I\}$ is a family of compact spaces, then $\underset{\alpha \in I}{\Pi} S_\alpha$ with the product topology τ_Π is a compact space.

Definition: If (S, τ) is a completely regular Hausdorff space, let $C = C_b(S, \mathbb{R})$, the set of all continuous and bounded functions $f : S \to \mathbb{R}$. For each $f \in C$, define a closed bounded (hence compact) interval in $\mathbb{R}$ by

$$I_f = [\inf_{x \in S} f(x), \sup_{x \in S} f(x)].$$

Define a map $e : S \to \underset{f \in C}{\Pi} I_f$ (with the product topology) by

$$e(x) = (f(x))_{f \in C}, \quad x \in S.$$

Then $\beta S = \overline{e(S)}$ is a compact space, called the **Stone-Čech compactification** of S.

Theorem A.1.10. (Stone-Čech) *Let S be a completely regular Hausdorff space and βS its Stone-Čech compactification. Let T be a compact Hausdorff space or $T = \mathbb{R}$. Then every continuous function $f : S \to T$ has a unique continuous extension $\widehat{f} : \beta S \to T$.*

Recall that, if S is a topological space, then a function $\varphi : S \to \mathbb{R}$ is *continuous* at $x_0 \in S$ if, for any $\varepsilon > 0$, there exists an open neighborhood $N(x_0)$ of x_0 such that

$$|\varphi(y) - \varphi(x_0)| < \varepsilon \text{ for all } y \in N(x_0),$$

or equivalently,

$$\varphi(x_0) - \varepsilon < \varphi(y) < \varphi(x_0) + \varepsilon \text{ for all } y \in N(x_0). \tag{*}$$

Definition: If S is a topological space, then a function $\varphi : S \to \mathbb{R}$ is said to be

(a) **upper semicontinuous** at $x_0 \in S$ if, for any $r \in \mathbb{R}$ with $\varphi(x_0) < r$, there exists an open neighborhood $N(x_0)$ of x_0 such that

$$\varphi(y) < r \text{ for all } y \in N(x_0);$$

(b) **lower semicontinuous** at $x_0 \in S$ if, for any $r \in \mathbb{R}$ with $\varphi(x_0) > r$, there exists an open neighborhood $N(x_0)$ of x_0 such that

$$\varphi(y) > r \text{ for all } y \in N(x_0);$$

(c) **upper** (resp. **lower**) **semicontinuous** on S if it is upper (resp. lower) semicontinuous at each point of S, or equivalently, if, for each $r \in \mathbb{R}$, the set $\{x \in S : \varphi(x) < r\}$ (resp. $\{x \in S : \varphi(x) > r\}$) is open (**[Will70]**, **p. 49**).

Clearly, by $(*)$, a function $\varphi : S \to \mathbb{R}$ is continuous at $x_0 \in S$ iff it is both upper and lower semicontinuous at $x_0 \in S$.

Theorem A.1.11. *(a) A function $\varphi : S \to \mathbb{R}$ is continuous iff it is both upper and lower semicontinuous.*

(b) The characteristic function χ_A of a subset A of S is upper (resp. lower) semicontinuous iff A is closed (resp. open).

(c) Every upper (resp. lower) semicontinuous function assumes its supremum (resp. infimum) on a compact set. In particular, every non-negative upper semicontinuous function on a compact set is bounded.

Definition: (**[Rud87]**, **p. 26**) Let $S_0 = \mathbb{N}$ and$\{S_n : n \geq 1\}$ be an sequence of infinite sets with $S_0 \supseteq S_1 \supseteq .... \supseteq S_n \supseteq ...$, and let r_n be the nth term of S_n (with respect to the natural order of the positive integers) and put

$$S = \{r_1, r_2, r_3, ...\}$$

For each $n \geq 1$, there are then at most $n - 1$ terms of S that are not in S_n. This construction of S from $\{S_n\}$ is the so-called **diagonal process.**

A.2   Banach Spaces

Definition. A set X with two operations of addition $(x, y) \to x + y$ and scalar multiplication $(\lambda, x) \to \lambda x$. is called a **vector space** over the field $\mathbb{K}$ $(= \mathbb{R}$ or $\mathbb{C})$ if the following condiond hold:

(a) X is an abelian group under addition,

(b) For any $x, y \in X$ and $\lambda, \mu \in \mathbb{K}$, the scalar multiplication satisfies:

(i) $\lambda(x + y) = \lambda x + \lambda y$

(iii) $(\lambda + \mu)x = \lambda x + \mu x$,

(ii) $(\lambda \mu)x = \lambda(\mu x)$,

(iv) $1.x = x$, where 1 is the multiplicative identity of $\mathbb{K}$.

Definition. A vector space A over a field $\mathbb{K}$ $(= \mathbb{R}$ or $\mathbb{C}$) is said to be an **algebra** if, for all $x, y \in A$, the product xy is defined and $xy \in A$, satisfying the following conditions:

(i) $x(yz) = (xy)z = xyz$,

(ii) $x(y + z) = xy + xz;\ (y + z)x = yx + zx$,

(iii) $(\lambda x)(\mu y) = (\lambda \mu)xy$,

where $\lambda, \mu \in \mathbb{K}$ and $x, y, z \in A$.

An algebra A is called a **commutative algebra** if $xy = yx$ for all $x, y \in A$. An element $e \in A$ is called an *identity* of A if $ex = xe = x$ for all $x \in A$; in this case, A is called an **algebra with identity**.

Definition. Let X be a vector space over $\mathbb{K}$. Then a function $\|.\| : X \to \mathbb{R}$ is called a *norm* on X if

(N_1) $\|x\| \geq 0$ for all $x \in X$;

(N_2) $\|x\| = 0 \Leftrightarrow x = 0$;

(N_3) $\|\lambda x\| = |\lambda|\,\|x\|$ for all $x \in X$ and $\lambda \in \mathbb{K}$;

(N_4) $\|x + y\| \leq \|x\| + \|y\|$ for all $x, y \in X$.

A vector space X with a norm $\|.\|$ is called a **normed vector space** or, simply, a **normed space**. Clearly, every normed space $(X, \|.\|)$ is a metric space with metric given by:

$$d(x, y) = \|x - y\|, \quad x, y \in X.$$

A normed space X is called a **Banach space** *if it is* complete with respect to the metric induced by the norm (i.e. if every Cauchy sequence in X converges to an element in X)

Definition. An algebra A with a norm $\|.\|$ is called a **normed algebra** if

$$\|xy\| \leq \|x\|\,\|y\| \ \text{ for all } \ x, y \in A.$$

A complete normed algebra A is called a **Banach algebra**.

Example 1. For any topological space S and X a topological vector space over $\mathbb{K}$, let $C(S, X)$ be the set of all continuous functions $f : S \to X$ and $C_b(S, X)$ the subset of $C(S, X)$ consisting of those functions which are bounded. Then both $C(S, X)$ and $C_b(S, X)$ are vector spaces over $\mathbb{K}$ under addition $f + g$ and scalar multiplication λf defined pointwise as:

$$(f + g)(x) = f(x) + g(x), (\lambda f)(x) = \lambda f(x), \quad x \in S.$$

If $X = \mathbb{K}$, then $C(S, \mathbb{K})$ and $C_b(S, \mathbb{K})$ are denoted by $C(S)$ and $C_b(S)$, repectively.

(a) If X is an algebra, then both $C(S, X)$ and $C_b(S, X)$ are also algebras with respect to the pointwise multiplication defined by

$$(fg)(x) = f(x)g(x), \quad x \in S.$$

If X is commutative, then $C(S, X)$ and $C_b(S, X)$ are also commutative In particular, $C(S)$ and $C_b(S)$ are commutative algebras.

(b) If X is only a vector space, then $C(S, X)$ is a $C(S)$-bimodule with respect to the module multiplications $(\varphi, f) \rightarrow \varphi.f$ and $(f, \varphi) \rightarrow f.\varphi$ defined by

$$(\varphi.f)(x) = \varphi(x)f(x) \text{ and } (f.\varphi)(x) = f(x)\varphi(x), \qquad x \in S.$$

Similarly, $C_b(S, X)$ is a $C_b(S)$-bimodule.

(c) If X is a normed space (resp. normed algebra), then $C_b(S, X)$ is also a normed space (resp. normed algebra) with sup norm $\|.\|$ defined by

$$\|f\|_\infty = \sup\{|\, f(x)\, |: x \in X\}, \qquad f \in C_b(S, X).$$

In particular, $C_b(S)$ is a normed algebra with identity $\mathbf{1}$, where $\mathbf{1}(x) = 1$ for all $x \in S$.

Note that $C_b(S, X)$ is a metric space with metric induced by the norm:

$$d(f, g) = \|f - g\|_\infty, \quad f, g \in C_b(S, X).$$

The topology induced by metric d on $C_b(S, X)$ is called the *uniform topology* and is denoted by u.

The above norm is not well-defined on $C(S, X)$ or $C(S)$ since $\|f\|_\infty$ does not exist for unbounded $f \in C(S, X)$ or $C(S)$. However, $C(S, X)$ is a metic space with respect to the metric d^* defined as

$$d^*(f, g) = \min\{1, d(f, g)\} = \min\{1, \|f - g\|_\infty\}, \quad f, g \in C(S).$$

Continuous Linear Mappings on Normed Spaces

Recall that, if $(X, \|\cdot\|_X)$ and $(Y, \|\cdot\|_Y)$ are normed spaces, then a linear mapping $T : X \rightarrow Y$ is **continuous** at $x_0 \in X$.if, given any $\varepsilon > 0$, there exists a $\delta > 0$ such that

$$\|T(x) - T(x_{0)}\|_Y < \varepsilon \text{ when ever } x \in X \text{ with } \|x - x_0\|_X < \delta.$$

For simplicity of notation we shall denote $(X, \|\cdot\|_X)$ and $(Y, \|\cdot\|_Y)$ by $(X, \|\cdot\|)$ and $(Y, \|\cdot\|)$. In the sequel, we always assume that X and Y, as vector spaces, are the same field $\mathbb{K}$ $(= \mathbb{R}$ or $\mathbb{C})$.

Definition. Let X and Y are normed spaces. Then a linear mapping $T : X \rightarrow Y$ is called **bounded** if it maps bounded sets of X into bounded sets of Y, or equivalently, there exists a constant $M > 0$ such that

$$\|T(x)\| \leq M.\|x\| \text{ for all } x \in X.$$

Theorem A.2.1. *Let X and Y be normed spaces (or TVSs). Then, for a linear mapping $T : X \rightarrow Y$, TFAE:*

(a) T is continuous at any one fixed point of X.

(b) T is is continuous on X

(c) T is bounded, i.e. there exists a constant $M > 0$ such that

$$\|T(x)\| \leq M.\|x\| \text{ for all } x \in X.$$

Thus, T is continuous on X iff T is continuous at $0 \in X$

Proof. $(a) \Rightarrow (b)$ Suppose T is continuous at any one fixed point $x_0 \in X$. Take any other point $x_1 \in X$, and let $\varepsilon > 0$. By continuity of T at $x_0 \in X$, there exists a $\delta > 0$ such that

$$||T(x) - T(x_{0)}|| < 1 \text{ whenever } x \in X \text{ with } ||x - x_0|| < \delta. \tag{1}$$

Now, let $x \in X$ with $||x - x_1|| < \delta$. Then, if $y = x - x_1 + x_0$,

$$||y - x_0|| = ||x - x_1 + x_0 - x_0|| = ||x - x_1|| < \delta,$$

and so by (1),

$$
\begin{aligned}
||T(y) - T(x_{0)}|| \; &< \; \varepsilon, \\
\text{or, } ||T(x) - T(x_1) + T(x_0) - T(x_{0)}|| \; &< \; \varepsilon \text{ (by linearity of } T) \\
\text{or, } ||T(x) - T(x_1)|| \; &< \; \varepsilon.
\end{aligned}
$$

That is,

$$||T(x) - T(x_{1)}|| < 1 \text{ whenever } x \in X \text{ with } ||x - x_1|| < \delta.$$

This shows that T is continuous at $x_1 \in X$. Since $x_1 \in X$ was arbitrarily chosen, T is continuous at all point of X.

$(b) \Rightarrow (c)$ Suppose T is continuous on X. Then since T is continuous at $0 \in X$, taking $\varepsilon = 1$, there exists a $\delta > 0$ such that

$$
\begin{aligned}
||T(x) - T(0)|| \; &< \; 1 \text{ whenever } x \in X \text{ with } ||x - 0|| < \delta, \\
\text{or, } ||T(x)|| \; &< \; 1 \text{ whenever } x \in X \text{ with } ||x|| < \delta. \tag{2}
\end{aligned}
$$

Now, for any $x(\neq 0) \in X$, let $z = \frac{\delta}{2||x||}$. Then

$$||z|| = ||\frac{\delta}{2||x||}x|| = \frac{\delta}{2||x||} \cdot ||x|| = \frac{\delta}{2} < \delta,$$

and so by (2),

$$||T(\frac{\delta}{2||x||}x)|| < 1, \text{ or } ||T(x)|| < \frac{2}{\delta}||x||.$$

Thus, if $M = \frac{2}{\delta}$, we have

$$||T(x)|| \leq M. ||x|| \text{ for all } x(\neq 0) \in X.$$

This clear hold for $x = 0$ also.

$(c) \Rightarrow (a)$ Suppose there exists a constant $M > 0$ such that

$$||T(x)|| \leq M.||x|| \text{ for all } x \in X. \tag{3}$$

Take any point $x_0 \in X$. Then by (3),

$$||T(x - x_0)|| \leq M.||x - x_0|| \ \text{for all} \ x \in X. \tag{4}$$

Let $\delta = \frac{\varepsilon}{2M}$. Then, if $||x - x_0|| < \delta$, by(4), we have

$$||T(x) - T(x_0)|| \leq M.\delta = M.\frac{\varepsilon}{2M} = \frac{\varepsilon}{2} < \varepsilon.$$

Thus T is continuous at $x_0 \in X$. $\square$

Note. In particular, $|| \cdot ||_1$ and$|| \cdot ||_2$ are two norms on a vector space X, the the identity (linear) mapping $I : (X, || \cdot ||_1)$ and $(X, || \cdot ||_2)$ is continuous on X iff there exists a constant $M > 0$ such that $||I(x)||_2 \leq M.||x||_1$ for all $x \in X$; that is

$$||x||_2 \leq M.||x||_1 \ \ \text{for all} \ \ x \in X.$$

Remark. Recall that if X is a topological space and Y a normed or metric space, then a function $f : X \to Y$ is said to be **bounded** if $f(X)$ is a bounded subset of Y. If X and Y are normed spaces and $T : X \to Y$ is a bounded linear mapping, then it is easy to see that $T(X)$ is a bounded subset of $Y \iff$ either $T = 0$ or $X = \{0\}$.

Theorem A.2.2. *Let X and Y be normed spaces and $T : X \to Y$. a linear mapping. Then T is bounded $\iff$ there exists a constant $M > 0$ such that*

$$||T(x)||_Y \leq M.||x||_X \ \text{for all} \ x \in X.$$

Definition. Let $(X, ||.||)$ and $(Y, ||.||)$ be a normed spaces. For any $T \in CL(X, Y)$, define

$$||T|| = \sup \ \{||T(x)||_Y : x \in X, ||x||_X \leq 1\}.$$

For simplicity of notation, we shall write shall $||.||$ fot both $||.||_X$ and $||.||_Y$.

Some useful properties of $(CL(X, Y), ||.||)$ are given, as follows.

Theorem A.2.3. *Let $(X, ||.||)$ and $(Y, ||.||)$ be normed spaces and $CL(X, Y)$ the vector space of all continuous linear mappings $T : X \to Y$. Then:*

(a) *$CL(X, Y)$ is a normed space with respect to the pointwise addition and scalar multiplication and the operator norm:*

$$||T|| = \sup\{||T(x)|| : x \in X, ||x|| \leq 1\}, T \in CL(X, Y).$$

(b) *$||T(x)|| \leq ||T|| \ ||x||$ for all $T \in CL(X, Y)$ and $x \in X$.*

(c) *If $Y = X$, then $BL(X) = BL(X, X)$ is a normed algebra with composition multiplication:*

$$(ST)(x) = S(T(x)), \qquad S, T \in BL(X),$$

and the norm as define above satisfies

$$||ST|| \leq ||S||.||T||, \qquad S, T \in BL(X),$$

Proof. (a) This is eacy to verify.

(b) Let $T \in CL(X,Y)$. Then, for any $x \in X$ with $x \neq 0$, $\left\|\frac{x}{\|x\|}\right\| = \frac{\|x\|}{\|x\|} = 1$ and so

$$\left\|T\left(\frac{x}{\|x\|}\right)\right\| \leq \sup_{\|y\|\leq 1} \|T(y)\| = \|T\|.$$

or

$$\frac{\|Tx\|}{\|x\|} \leq \|T\| \quad \text{or} \quad \|Tx\| \leq \|T\|.\|x\|.$$

If $x = 0$, clealy

$$\|T(x)\| = \|T(0)\| = 0 = \|T\| .\|0\| = \|T\|.\|x\|.$$

(c) Let $S, T \in CL(A)$. Using (b),

$$\begin{aligned}
\|ST\| &= \sup_{\|x)\leq 1} \|S(T(x))\| \leq \sup_{\|x)\leq 1} \|S\|.\|T(x)\| \\
&\leq \sup_{\|x)\leq 1} \|S\|.\|T\|.\|x\| = \|S\|.\|T\|.
\end{aligned}$$

Therefore $\|.\|$ is submultiplicative norm on $CL(X)$. Thus $(CL(X), \|.\|)$ is a normed algebra. $\square$

Definition. Let $(X, \|.\|)$ and $(Y, \|.\|)$ be normed algebras.

(1) The **uniform operator topology** u on $CL(X,Y)$ is given by the norm

$$\|T\| = \sup \{\|T(x)\| : x \in X, \|x\|_X \leq 1\}.$$

(2) The **strong operator topology** s on $CL(X,Y)$ is the locally convex topology generated by the family $\{\|.\|_a : a \in X\}$ of seminorms, where

$$\|T\|_a = \|T(a)\|, \quad T \in CL(X,Y).$$

Clearly, $s \leq u$.

Theorem A.2.5. [DS58, Ii.3.8, p. 61] Let $(X, \|.\|)$ and $(Y, \|.\|)$ be normed spaces. *If Y is complete, then $(CL(X,Y), \|.\|)$ is also complete.*

Proof. Let $\{T_n\}$ be a Cauchy sequence in $(CL(X,Y), \|.\|)$. We show that there exists a $T \in CL(X,Y)$ such that $T_n \xrightarrow{\|.\|} T$.

Step (I). For each $x \in X$, $\{T_n(x)\}$ is a Cauchy sequence in Y, as follows. [This clearly holds for $x = 0$. Let $x \in X$ with $x \neq 0$, and let $\varepsilon > 0$. Since $\{T_n\}$ is a Cauchy sequence in $(CL(X,Y), \|.\|)$, there exists an integer $N \geq 1$ such that

$$\|T_n - T_m\| \quad < \quad \frac{\varepsilon}{\|x\|} \quad \text{for all} \quad n, m \geq N;$$

hence $\|T_n(x) - T_m(x)\| \leq \|T_n - T_m\|.\|x\| < \frac{\varepsilon}{\|x\|}\|x\| = \varepsilon$ for all $n, m \geq N$.

So $\{T_n(x)\}$ is a Cauchy sequence in Y.] Since Y is complete, the mapping $T : X \longrightarrow Y$ given by

$$T(x) = \lim_n T_n(x), \ x \in X,$$

is well-defined.

Step (II). To show $T \in CL(X,Y)$: Clearly, T is linear. We next show that T is continuous, or equivalently that T is bounded. Since $\{T_n\}$, being a Cauchy sequence in $(CL(X,Y), ||.||)$, is bounded, there exists a constant $M > 0$ such that

$$||T_n|| \leq M \ \text{ for all } \ n \geq 1.$$

Hence

$$||T_n(x)|| \leq ||T_n||.||x|| \leq M.||x|| \ \text{ for all } \ x \in X \text{ and } \ n \geq 1.$$

Therefore, by continuity of norm,

$$||T(x)|| = ||\lim_{n \to \infty} T_n(x)|| = \lim_{n \to \infty} ||T_n(x)|| \leq \sup_n \quad ||T_n||.||x||(x)| \leq M.||x||$$
$$\text{for all } \ x \in X.$$

Thus $T \in CL(X,Y)$.

Step (III). To show that $T_n \xrightarrow{||.||} T$, let $x \in X$ and let $\varepsilon > 0$. Since $\{T_n\}$ is a Cauchy sequence in $CL(X,Y)$, there exists an integer $k \geq 1$ such that

$$||T_n - T_m|| \ < \ \frac{\varepsilon}{2} \ \text{ for all } \ n, m \geq k,$$

or $\quad ||T_n(x) - T_m(x)|| \ \leq \ ||T_n - T_m||.||x|| < \frac{\varepsilon}{2}||x|| \ \text{ for all } \ x \in X, \ n, m \geq k.$

Fix $n \geq k$ and let $m \to \infty$, we have

$$||T_n(x) - T(x)|| \leq \frac{\varepsilon}{2}||x|| < \varepsilon||x|| \ \text{ for all } \ x \in X.$$

Hence

$$||T_n - T|| = \sup_{||x|| \leq 1} ||T_n(x) - T(x)|| \leq \sup_{||x|| \leq 1} \frac{\varepsilon}{2}||x|| = \frac{\varepsilon}{2} < \varepsilon \ \text{ for all } \ n \geq k.$$

Thus $T_n \xrightarrow{||.||} T$. Hence $(CL(X,Y), ||.||)$ is complete. $\square$

Definition: Let X be a TVS or a normed space. Then any linear map $f : X \to \mathbb{K}$ is called a *linear functional* on X. The set of all continuous linear functionals on X is also a vector space, called the *topological dual* of X and is denoted by X^*. Clearly, $X^* = CL(X, \mathbb{K})$.

Definition: Let X and Y be topological spaces and $T : X \to Y$ a mapping. Then:

(a) The set $G(T) = \{(x, Tx) : x \in X\}$ is called the **graph** of T in $X \times Y$.

(b) T is said to have a **closed graph** if $G(T)$ is a closed subset of $X \times Y$. (Here $X \times Y$ has the product topology.)

The following theorem gives equivalent condition for a mapping $T : X \rightarrow Y$ to have a closed graph.

Theorem A.2.5. *Let X and Y be topological spaces and $T : X \times Y$ a mapping. Then the following are equivalent*

(i) T has a closed graph.

(ii) If $\{x_\alpha\}$ is a net in X with $x_\alpha \rightarrow x \in X$ and $Tx_\alpha \rightarrow y \in Y$, then $y = T(x)$

Proof. (i) $\Rightarrow$ (ii). Let $\{x_\alpha = \alpha \in I\} \subseteq X$ with $x_\alpha \rightarrow x \in X$ and $Tx_\alpha \rightarrow y \in Y$. Then clearly $(x_\alpha, Tx_\alpha) \rightarrow (x, y)$ in $X \times Y$ (in the product topology). Since $\{(x_\alpha, Tx_\alpha) = \alpha \in I\} \subseteq G(T)$ and, by (i), $G(T)$ is closed, we have $(x, y) \in G(T)$. Hence $y = Tx$.

(ii) $\Rightarrow$ (i). Suppose (ii) holds. To show that $\overline{G(T)} \subseteq G(T)$, let $(x, y) \in X \times Y$ with $(x, y) \in \overline{G(T)}$. Then there exists a net $\{(x_\alpha, Tx_\alpha) : \alpha \in I\} \subseteq G(T)$ such that $(x_\alpha, Tx_\alpha) \rightarrow (x, y)$. This implies that $x_\alpha \rightarrow x$ and $Tx_\alpha \rightarrow y$. Hence, by hypothesis (ii), $y = Tx$. So $(x, y) = (x, Tx) \in G(T)$. Thus $\overline{G(T)} = G(T)$ and so T has a closed graph. $\square$

The next result is concerned with the questions: Under what conditions continuity of T implies that T has a closed graph ? Conversely, under what conditions, T has a closed graph implies that T is continuous ?

Theorem A.2.6. *Let X and Y be topological spaces with Y Hausdorff, and let $T : X \rightarrow Y$ be a mapping. If T is continuous, then T has a closed graph.*

Proof. Suppose T is continuous. To show that T has a closed graph, let $(x, y) \in X \times Y$ with $(x, y) \in \overline{G(T)}$. Then there exists a net $\{(x_\alpha, Tx_\alpha) : \alpha \in I\} \subseteq G(T)$ such that $(x_\alpha, Tx_\alpha) \rightarrow (x, y)$. This implies that $x_\alpha \rightarrow x$ and $Tx_\alpha \rightarrow y$. Since T is continuous and $x_\alpha \rightarrow x$, we have $Tx_\alpha \rightarrow Tx$ in Y. Since Y is Hausdorff, $\{Tx_\alpha\}$ has a unique limit and so $y = Tx$. Hence $(x, y) = (x, Tx) \in G(T)$. Thus $\overline{G(T)} = G(T)$; i.e. T has a closed graph. $\square$

We now state some fundamental theorems of Functional Analysis for Banach spaces.

Theorem A.2.7. (Closed graph theorem) [**Sim, p. 238**] *Let X and Y be Banach spaces. If $T : X \rightarrow Y$ is linear and has closed graph, then T is continuous. Hence a linear map $T : X \rightarrow Y$ is continuous iff T has has closed graph.*

Theorem A.2.8. (Open mapping theorem) [**Sim, p. 236**] *Let X and Y be Banach spaces. If $T : X \rightarrow Y$ is a continuous and onto linear mapping, then T is an open mapping (i.e., T maps open sets into open sets). If, in addition T is one-one, then $T^{-1} : Y \rightarrow X$ is also continuous; hence T is a homeomorphism.*

Theorem A.2.9.(Uniform boundedness principle) [**Sim, p. 240-241**] *Let X be a Banach space and Y any normed space. If a collection $\mathcal{F} \subseteq BL(X, Y)$ is pointwise bounded (i.e. for each $x \in X$, the set $\mathcal{F}(x) = \{T(x) : T \in \mathcal{F}\}$ is bounded in Y), then $\mathcal{F}$ is uniformly bounded in $BL(X, Y)$ (i.e. $\{\|T\| : T \in \mathcal{F}\}$ is a bounded set in $\mathbb{R}$).*

Theorem A.2.10. (Banach-Steinhaus) [**DS58, p. 60; Sim, p. 242**]

Let X be a Banach space and Y any normed space. Let $\{T_n\}$ be a sequence in $BL(X, Y)$ such that, for each $x \in X, T(x) = \lim_{n \to \infty} T_n(x)$ exists. Then T is continuous on X.

Theorem A.2.11. (Hahn-Banach theorem for Normed Spaces) **[Sch71, p. 49]** *Let $(X, \|\cdot\|)$ be a normed space, M a vector subspace of X and f a continuous linear functional on M. Then there exists a $g \in X^*$ (called the extension of f from M to X) such that*

$$g \;=\; f \text{ on } M \text{ and } \|g\|_X = \|f\|_M\,;$$
$$\text{(i.e. } \sup\{|g(x)| \;:\; x \in X, \|x\| \leq 1\} = \sup\{|f(y)| : y \in M, \|y\| \leq 1\})$$

Corollary A.2.12. [DS58, II.3.14] *Let $(X, \|\cdot\|)$ be a normed space. Then, for any $x_0 \neq 0$ in X, there exists a $g \in X^*$ such that*

$$g(x_0) = \|x_0\| \text{ and } \|g\| = 1.$$

In particular, X^* separates points of X (i.e. for any $x \neq 0$ in X, there exist a $g \in X^*$ such that $g(x) \neq 0$).

Definition: [Sim63, p. 233] (1) Let $(X, \|\cdot\|)$ be an normed space. For any $x \in X$, we define the **evaluation map** $\widehat{x} : X^* \longrightarrow \mathbb{K}$ by

$$\widehat{x}(f) = f(x), \quad f \in X^*.$$

Then $\widehat{x}$ is linear and $|\widehat{x}(f)| = |f(x)| \leq \|f\|.\|x\|$ for all $f \in X^*$, and so $\widehat{x} \in X^{**} = (X^*)^*$.

Corollary A.2.13. *Let $(X, \|\cdot\|)$ be a normed space. Then, for any $x \in X$,*

$$\|x\| = \|\widehat{x}\| = \sup\{\|f(x) : f \in X^*, \|f\| \leq 1).$$

Definition: We define the **embedding map** $\pi : X \longrightarrow X^{**}$ by

$$\pi(x) = \widehat{x}, \quad x \in X.$$

Clearly, $\widehat{x} = 0$ implies that $x = 0$ and so π is one-one. Hence we may identify X with a subspace $\widehat{X} = \{\widehat{x} : x \in X\}$ of X^{**}, so $\widehat{X} \subseteq X^{**} \subseteq (X^*)^*$. Further, the map $\pi : X \longrightarrow \pi(X) = \widehat{X}$ is a homeomorphism.

Definition: Let $(X, \|\cdot\|)$ be an normed space. If $\pi : X \longrightarrow X^{**}$ is onto (i.e. if $\widehat{X} = X^{**}$), then X is called **reflexive.**

Note. If $(X, \|\cdot\|)$ is a reflexive normed space, then, since $\pi : X \longrightarrow X^{**}$ is an onto isometry and $X^{**} = L(X^*, K)$ is complete, it follows that $(X, \|\cdot\|)$ is complete and hence a Banach space.

Adjoints of Linear Mappings

Definition: [BD73], p. 140; KN76], p. 199] Let X and Y be normed spaces with X^* and Y^* as dual spaces, let $T : X \longrightarrow Y$ a linear map. Then a linear map $T^* : Y^* \longrightarrow X^*$ is called the *adjoint (or conjugate, or transpose)* of T if

$$T^*(g)(x) = g(Tx) \ \text{ for all } g \in Y^* \text{ and } x \in X;$$

that is, $T^*(g) = g \circ T$ for all $g \in Y^*$.

Similarly, $T^{**} : X^{**} \longrightarrow Y^{**}$ is given by

$$T^{**}(F)(g) = Y(T^*g) \ \text{ for all } F \in X^{**} \text{ and } g \in F^*;$$

that is, $T^{**}(F) = F \circ T^*$ for all $F \in X^{**}$.

Remarks. (1) If $S, T \in CL(X,Y)$ and $\alpha \in \mathbb{K}$, then

$$
\begin{aligned}
(S+T)^* &= S^* + T^* \\
(\alpha T)^* &= \overline{\alpha} T^*.
\end{aligned}
$$

(2) If $I : X \longrightarrow X$ is the identity map, then $I^* : X^* \longrightarrow X^*$ is also the identity map.

(3) If X, Y, Z are normed spaces and $T \in CL(X,Y)$ and $S \in CL(Y,Z)$, then

$$(ST)^* = T^* S^*.$$

(4) If $T \in CL(X,Y))$, then

$$(T^*)^n = (T^n)^*.$$

Theorem A.2.14. *Let X and Y be Banach spaces and $T : X \longrightarrow Y$ a continuous linear mapping. Then:*

(i) *If T is onto, then T^{**} is also onto.*

(ii) *If T^{**} is onto and $T(X)$ is dense in Y, then T is onto.*

Theorem A.2.15 *Let X and Y be normed spaces. Then*

(i) *if $T \in CL(X,Y)$, then $T^* \in CL(Y^*, X^*)$ and*

$$\|T^*\| = \|T\| \,;$$

hence $T^{**} \in CL(X^{**}, Y^{**})$ and

$$\|T^{**}\| = \|T\| \,.$$

(ii) *if $T \in CL(X,Y)$ and T^{-1} exists with $T^{-1} \in CL(Y,X)$, then $(T^*)^{-1}$ also exists, $(T^*)^{-1} \in CL(X^*, Y^*)$ and*

$$(T^*)^{-1} = (T^{-1})^*.$$

Definition. Let X and Y be Banach spaces. A bounded linear mapping $T : X \to Y$ is said to be **weakly compact** if for any bounded set B in X, the weak closure of TB is compact in the weak topology of Y.

Remarks. Let us note the following facts about weakly compact operators [**DS58, Chap.VI**].

(1) A bounded linear mapping $T : X \to Y$ is weakly compact if and only if $T^{**}X^{**}$ is contained in the natural embedding $\gamma(Y)$ of Y into Y^{**}.

(2) A bounded linear mapping T is weakly compact if and only if its adjoint T^* is weakly compact.

Definition. Let A be an algebra and Y a vector space over the same field $\mathbb{K}$. Then Y is called a **left A-module** if there exists a map $(a, x) \to a.x$ of $A \times Y \to Y$, called **module multiplication**, such that, for all $a, b \in A$ and $x, y \in Y$,

(1) $a \in A$ and $x \in X \Rightarrow a.x \in Y$ (i.e. $A.Y \subseteq Y$),

(2) $(a + b).x = a.x + b.x$,

(3) $a.(x + y) = a.x + a.y$,

(4) $(ab).x = a.(b.x)$,

(5) $X.x = x$ if A has an identity X (and in this case Y is called a **unitary** module).

Similarly, Y is called a **right A-module** if we consider the module multiplication map as $(x, a) \to x.a$ of $Y \times A \to Y$ (i.e. $Y.A \subseteq Y$) which satisfies the corresponding right module multiplication properties. Y is called an **A-bimodule** (or simpy an *A-module)* if it is both a left A-module and a right A-module. Clearly, an algebra A is an A-bimodule.

A.3   Topological Vector Spaces

Definition: If X is a vector space and τ is a topology on X, then the pair (X, τ) is called a **topological vector space** *(**TVS**, in short)* if the following conditions hold:

(TVS$_1$) The operation of addition $(x, y) \to x + y$ of $X \times X \to X$ is jointly continuous; i.e., given any $x, y \in X$ and any τneighborhood U of $x + y$ in X, there exist τ-neighborhoods V of x and W of y in X such that $V + W \subseteq U$.

(TVS$_2$) The operation of scalar multiplication $(\lambda, x) \to \lambda x$ of $\mathbb{K} \times X \to X$ is jointly continuous; i.e., given any $x \in X$ and $\lambda \in \mathbb{K}$ and any τneighborhood U of λx in X, there exist a $\tau-$neighborhood V of x in X and a neighborhood D of λ in $\mathbb{K}$ such that $DV \subseteq U$.

Defonition: A function $p : X \to \mathbb{R}$ is called a **seminorm** on X if it satisfies

(S$_1$) $p(x) \geq 0$ for all $x \in X$;

(S$_2$) $p(x) = 0$ if $x = 0$;

(S$_3$) $p(\lambda x) = |\lambda| p(x)$ for all $x \in X$ and $\lambda \in \mathbb{K}$ *(absolutely homogeneous)*;

(S$_4$) $p(x + y) \leq p(x) + p(y)$ for all $x, y \in X$ *(subadditive)*.

If a seminorm p satisfies $p(x) = 0$ implies $x = 0$, then p is called a **norm** on X.

Remarks. (1) If p is a seminorm on a vector space X, it may happen that $p(x) = 0$ for some $x \neq 0$. Hence, if p is a seminorm on a vector space X, then it defines a pseudometric d_p on X given by

$$d_p(x, y) = p(x - y), \quad x, y \in X.$$

Definition: A TVS (X, τ) is called:

(i) a **locally convex space** (or a **locally convex TVS**) if it has a base of convex neighbourhoods of 0;

(ii) **metrizable** if τ is induced by a metric d on X.

Theorem A.3.1. *If (X, τ) is a locally convex space, then its topology τ can be defined by a family of seminorms. Conversely, if $\mathcal{P}$ is a family of seminorms on a vector space X over $\mathbb{K}$, then $\mathcal{P}$ defines a locally convex topology τ (say) on X and a base of τ-neighborhoods of 0 consists of all sets of the form*

$$\{x \in X : \max_{1 \leq j \leq n} p_j(x) \leq \varepsilon\}, \text{, where } \varepsilon > 0 \text{ and } p_j \in \mathcal{P}, \, n \in \mathbb{N}.$$

Theorem A.3.2. *Let X and Y be TVSs. Then:*

(a) A linear mapping $T : X \to Y$ is continuous on X iff T is continuous at $0 \in X$ iff T is uniformly continuous on X.

(b) If $f : X \to K$ a non-zero linear functional, then f is continuous on X iff $f^{-1}(0)$ is closed in X iff f is bounded on some neighborhood U of 0 in X.

Definition: (1) The set of all linear functionals on a vector space (or a TVS) X is a vector space, called the **algebraic dual** of X and is denoted by X'.

(2) If X is a TVS, the set of all continuous linear functionals on X is also a vector space, called the **topological dual** of X and is denoted by X^*.

Clearly, X' is a vector space over $\mathbb{K}$. Further, X^* is a vector subspace of X'.

Theorem A.3.3. (Hahn-Banach theorem for TVSs) [**Ko1, p. 191; Sch71, p. 47**] *Let (X, τ) be a TVS, M a vector subspace of X, and p a continuous seminorm on X. If f is a continuous linear functional on M with $|f(x)| \leq p(x)$ for all $x \in M$, then there exists a continuous linear functional g on X (i.e. $g \in X^*$) such that*

$$g = f \text{ on } M \quad \text{and} \quad |g(x)| \leq p(x) \text{ for all } x \in X.$$

Theorem A.3.4. (Hahn-Banach theorem for LCSs) [Sch71, p. 49] *Let X be an LCS, M a vector subspace of X and f a continuous linear functional on M. Then there exist a $g \in X^*$ and a continuous seminorm p on X such that*

$$g = f \text{ on } M \text{ and } |g(x)| \leq p(x) \text{ for all } x \in X.$$

Corollary A.3.5. **[DS58, p64; Naim72, p. 54]** *Let X be a LCS, M a closed vector subspace of X and $x_0 \in X$ with $x_0 \notin M$. Then there exists a $g \in X^*$ such that $g = 0$ on M and $g(x_0) = 1$.*

Corollary A.3.6. **[Naim72, p. 54]** *Let X be a LCS and M a vector subspace of X. Then M is dense in $X \iff$ for any $g \in X^*$ with $g = 0$ on M, we have $g = 0$ on X.*

Corollary A.3.7. *Let (X, τ) be a TVS over $\mathbb{K}$, $x_0 \in X$, and p a continuous seminorm on X. Then there exist $g \in X^*$ such that $g(x_0) = p(x_0)$ and $|g(x)| \leq p(x)$ for all $x \in X$.*

Corollary A.3.8. *Let X be a Hausdorff LCS (in particular, a normed space) over $\mathbb{K}$. Then, for any $x(\neq 0) \in X$, there exist a $g \in X^*$ such that $g(x) \neq 0$. In particular, X^* separates points of X .*

Note that if X is a non-LCS, then X^* may or may not separate the points of X.

Definition. **[DS58, p. 52]** Let X and Y be TVSs, and let $CL(X, Y)$ denote the set of all continuous linear mappings from $X \to Y$. A subset H of $CL(X, Y)$ is said to be *equicontinuous* if for each neighborhood V of 0 in Y, there exists a neighborhood U of 0 in X such that $T(U) \subseteq V$ for all $T \in H$.

Theorem A.3.9. (Principle of uniform boundedness) [Rud91, p. 45] *Let X be a complete metrizable TVS and Y any TVS. Let $A = \{T_\alpha : \alpha \in I\}$ be a collection of continuous linear mappings from X into Y such that, for each $x \in X$, the set $A(x) = \{T_\alpha x : \alpha \in I\}$ is bounded in Y. Then A is equicontinuous; hence, for any bounded set B in X, $\bigcup\{T_\alpha(B) : \alpha \in I\}$ is a bounded set in Y.*

Theorem A.3.10. (Banach-Steinhaus) **[DS58, p. 55; Rud91,, p. 46]** *Let X be a complete metrizable TVS and Y any TVS. Let $\{T_n\}$ be a sequence of continuous linear mappings from X into Y such that, for each $x \in X$, the set*

$$T(x) = \lim_{n \to \infty} T_n(x)$$

exists. Then T is continuous on X.

Theorem A.3.11. (Open mapping theorem) **[Rud91, p. 45; Sch71, p. 77]** *Let X and Y be complete metrizable TVSs and $T : X \to Y$ a continuous and onto linear mapping. Then T is an open mapping. If, in addition T is one to one, then $T^{-1} : Y \to X$ is also continuous.*

Recall that if X and Y are TVSs, then a linear mapping $T : X \to Y$ is said to have a *closed graph* if its graph $G(T) = \{(x, Tx) : x \in X\}$ is closed in $X \times Y$ or, equivalently, for any net $\{x_\alpha\}$ in X with $x_\alpha \to x \in X$ and $T(x_\alpha) \to y \in Y$, we have $y = T(x)$.

Every continuous linear map $T : X \to Y$ has a closed graph. The converse holds under some additional hypotheses.

Theorem A.3.12. (Closed graph theorem) **[Rud91, p. 51; Sch71, p. 78]** *Let X and Y be complete metrizable TVSs. If $T : X \to Y$ is linear and has closed graph (i.e. the graph $G(T) = \{(x, Tx) : x \in X\}$ of T is* *closed in $X \times Y$), then T is continuous.*

Barrelled and Bornological Spaces

Definition. [Sch71, p. 60-61] Let X be a TVS. A subset D of X is called a

(a) **barrel** if it is closed, convex, balanced and absorbing;

(b) **bornivorous set** if it absorbs every bounded subset of X.

Definition. [Sch71, p. 60-61. 142] A LCS X is called a

(a) **barrelled space** if every barrel in X is a neighbourhood of 0;

(b) **bornological space** if every balanced convex bornivorous subset of X is a neighbourhood of 0 in X.

(c) **quasi-barrelled** (or *infrabarrelled*) if every bornivorous barrell in X is a neighbourhood of 0 in X.

Clearly, every barrelled space and every bornological space is quasibarrelled.

Note. [Sch71, p.63] In general, a barrelled space $\nLeftrightarrow$ a bornological space. Let

$$X \subset \{f \in C[0,1] : f = 0 \text{ in a neighbourhood (depending on } f) \text{ of } t = 0\}$$

with the sup-norm topology, and let $D = \{f \in X : n|f(\frac{1}{n})| \leq 1, n \in \mathbb{N}\}$. Then D is a barrelled in X but not a neighbourhood of 0 [Sch71, p. 70]. Thus we have a normed (hence bornological) space X which is not barrelled. Further, the classical examples of Nachbin and Shirota show that a barreled space need not be bornological

Theorem A.3.13. [Sch71, p. 60-61](i) *Every complete metrizable LCS is barrelled; in particular, every Banach space is barrelled.*

(ii) *Every metrizable LCS is bornological; in particular, every normed space is bornological.*

Theorem A.3.14. [Sch71, p. 63, 142] *Let (X, τ) be a quasi-complete (in particular, a complete) LCS. Then:*

(i) *X is quasi-barrelled $\implies X$ is barrelled (and hence $\iff$ holds).*

(ii) *X is bornological $\implies X$ is barrelled.*

Theorem A.3.15. [Sch71, p.62] *Let X be a bornological LCS and F any LCS. Then every bounded linear map $T : X \to F$ is continuous (i.e. T is continuous $\iff T$ is bounded).*

Definition A.3.16. [Edw65] A TVS X is called *ultrabornological* if every bounded linear map from X into any TVS is continuous. X is called *ultrabarrelled (quasi-ultrabarrelled)* if every closed (bounded) linear map from X into any complete metrizable TVS is continuous. We mention that every ultrabornological and every ultrabarrelled TVS is quasi-ultrabarrelled. Further, every metrizable TVS is ultrabornological.

A.4   Duality Theory and Weak Topologies

Duality Theory (via seminorms)

Definition. Let X and Y be vector spaces over $\mathbb{K}$. Then the pair (X, Y) is called a **dual pair** if there exists a bilinear map $<,>: X \times Y \to \mathbb{K}$ such that

(i) for each $x \neq 0$ in X, there exists a $g \in Y$ such that $< x, g > \neq 0$;

(ii) for each $g \neq 0$ in Y, there exists an $x \in X$ such that $< x, g > \neq 0$.

Clearly, if (X, Y) is a dual pair, then (Y, X) is also a dual pair.

If (X, τ) is a TVS such that X^* separates points of X. (e.g. if X is an LCS), then (X, X^*) is a dual pair. For convenience, we shall denote a dual pair by (X, X^*) instead of (X, Y) and write $g(x)$ for $< x, g >$.

Definition. *Let (X, X^*) be a dual pair.*

(a) For any $f \in X^*$, define $p_f : X \to \mathbb{R}$ by

$$p_f(x) = |f(x)| \text{ for all } x \in X,$$

Then $\{p_f : f \in X^*\}$ is a family of seminorms on X, and so this defines a locally convex topology on X. This topology is called **weak topology** on X and is denoted by $w = w(X, X^*)$ or $w(X, X^*)$.

(b) Similarly, for any $x \in X$, define $q_x : X^* \to \mathbb{R}$ by

$$q_x(f) = |f(x)| \text{ for all } f \in X^*.$$

Then $\{q_x : x \in X\}$ is a family of seminorms on X^*, and so this defines a locally convex topology on X^*. This topology is called the **weak* topology** on X^* denoted by $w^* = w^*(X^*, X) = w(X^*, X)$.

Theorem A.4.1. *(a) $(X, w(X, X^*))^* = (X, \tau)^*$.*

(b) (Mazur) For any convex subset M of X, $\overline{M}^w = \overline{M}^\tau$.

(c) (Mackey) For any subset M of X, M is τ-bounded iff M is $w(X, X^*)$-bounded.

(d) If (X, τ) is finite-dimensional, then $\tau = w(X, X^*)$.

We next consider strong and strong* topologies on X and X^*, respectively.

Definition. For any set $B \subseteq X$, the mapping $q_B : X^* \to \mathbb{R}$ defined by

$$q_B(f) = \sup\{|f(x)| : x \in B\}, \ f \in X^*,$$

is a seminorm on X^* iff B is $w(X^*, X)$-bounded in X^* (see **[RR64], p.45; Hor66, p. 195]**).

Definition: Let (X, τ) be a TVS. For any $w(X^*, X)$-bounded set $A \subseteq X^*$, the mapping $p_A : X \to \mathbb{R}$ defined by

$$p_A(x) = \sup\{|f(x)| : f \in A\}, \ x \in X,$$

is a seminorm p_A on X. If $\mathcal{A}$ is the collection of all $w(X^*, X)$-bounded subsets of X^*, then the collection $\{p_A : A \in \mathcal{A}\}$ of seminorms defines a locally convex topology on X, called the *strong topology* and denoted by $b = b(X, X^*)$. Clearly, $w(X, X^*) \leq b(X, X^*)$. Similarly, for any τ-bounded (or equivalently, $w(X^*, X)$-bounded) set $B \subseteq X$, the mapping $q_B : X^* \to \mathbb{R}$ defined by

$$q_B(f) = \sup\{|f(x)| : x \in B\}, \ f \in X^*.$$

is a seminorm q_B on X^*. If $\mathcal{B}$ is the collection of all τ-bounded subsets of X, then the collection $\{q_B : B \in \mathcal{B}\}$ of seminorms defines a locally convex topology on X^*, called the *strong* topology* and denoted by $b^* = b(X^*, X)$ (or by $b(X^*, X)$ in some books). Clearly, $w(X^*, X) \le b(X^*, X)$.

Definition. [Sch71, p. 130] Let (X, X^*) be a dual pair. A topology τ on X is called **compatible** (or **consistent**) with the dual pair if τ is locally convex topology and $(X, \tau)^* \equiv X^*$. In this case, τ is also called a **topology of the dual pair**.

Theorem A.4.2. [Sch71, p. 132] *Let (X, X^*) be a dual pair and τ a compatible topology on X. Then a subset M of X is τ-bounded $\iff M$ is $w(X, X^*)$-bounded.*

Theorem A.4.3. *Let (X, X^*) be a dual pair. Then $w(X, X^*)$ is compatible with the dual pair and is the smallest compatible topology for X.*

Definition. Let (X, X^*) be a dual pair. The largest compatible topology on X is the Mackey topology $m(X, X^*)$. Thus a locally convex topology τ on X is compatible $\iff w(X, X^*) \subseteq \tau \subseteq m(X, X^*)$.

Note. [Sch71, p. 143; Hor66, p. 209] In general, the strong*-topology $b(X^*, X)$ on X^* need not be compatible.

Remarks. (1) [Wila78, p. 119] If $(X, || \cdot ||)$ is a normed space, then the strong* topology $b(X^*, X)$ on X^* is the norm topology given by

$$||f|| = \sup\{|f(x)| : ||x|| \le 1, \ f \in X^*.$$

However, in this case, the strong topology $b(X, X^*)$ on X need not the norm topology for X; in general, $|| \cdot || \le b(X, X^*)$. If X is a Banach space, then $b(X, X^*) = || \cdot ||$.

(2) The weak and strong topologies are often useful in giving counter-examples. For instance, the Banach space $\ell_1 = \ell_1(\mathbb{N})$ with the weak topology $w = w(\ell_1, \ell_\infty)$ is sequentially complete but not complete; ℓ_1 with the weak* topology $w^* = w(\ell_1, c_o)$ is quasi-complete but not complete. (Recall that $c_o^* \cong \ell_1$ and $\ell_1^* = \ell_\infty$ as normed spaces.) Hilbert space is weakly quasi-complete but not weakly complete.

Duality Theory (via polars)

Definition. [Sch71, p125; Ko69, p.245; Hor66, p. 190] Let (X, X^*) be a dual pair. For each subset A of X, the set

$$A^\circ = \{f \in X^* : \sup_{x \in A} |f(x)| \le 1\}$$

is called the *polar* of A in X^*. The set

$$A^{\circ\circ} = \{x \in X : \sup_{f \in A^\circ} |f(x)| \le 1\}$$

is called the *bipolar* of A in X.

Theorem A.4.4. Let X be a TVS with non-trivial dual X^*, and let $A, B, A_\alpha \subseteq X$.

(a) $\varnothing^o = X^*$ and $X^o = \{0\}$

(b) $(\lambda A)^o = \frac{1}{\lambda} A^o$ for $\lambda(\neq 0) \in \mathbb{K}$

(c) If $A \subseteq B$, then $B^o \subseteq A^o$

(d) $A \subseteq A^{oo}$

(e) $A^o = A^{oo}$

(f) $(\bigcup_{\alpha \in I} A_\alpha)^o = \bigcap_{\alpha \in I} A_\alpha^o$

(g) If A is a vector subspace of X, then $A^o = \{g \in X^* :< x, g >= 0 \ \forall \ x \in A\}$.

Theorem A.4.5. [**Sch71, p. 126**] *Let* (X, X^*) *be a dual pair and* $A \subseteq X$. *Then:*

(i) A^o *is convex, balanced and* $w(X^*, X)$-*closed in* X^*.

(ii) A^o *is absorbing* $\Leftrightarrow$ A *is* $w(X, X^*)$-*bounded* $\Leftrightarrow$ $p_A(f) = \sup_{x \in A} |f(x)|$

defines a seminorm p *on* X^*.

Theorem A.4.6. (Bipoler therem) [**Sch71, p. 126**] *If* (X, τ) *is a LCS, then* A^{oo} *is equal to the* τ-*closed balanced convex hull of* A.

Definition. Let (X, X^*) be a dual pair and $\mathcal{S}_{wb} = \mathcal{S}_{wb}(X)$, the collection of all $w(X, X^*)$-bounded subsets of X. Fix any $\mathcal{S} \subseteq \mathcal{S}_{wb}$. The collection $\mathcal{S}^o = \{B^o : B \in \mathcal{S}\}$ consists of absorbing, balanced, convex and $w(X^*, X)$-closed subsets of X^* and so it generates a locally convex topology on X^*, called the *polar topology* on X^* associated with $\mathcal{S}$ and is denoted by $t_\mathcal{S}$.

Theorem A.4.7. *Let* (X, X^*) *be a dual pair and let* $\mathcal{S} \subseteq \mathcal{S}_{wb}$. *If* $X = \bigcup\{B : B \in \mathcal{S}\}$, *then* $t_\mathcal{S}$ *is Hausdorff.*

We now consider some important classes of $t_\mathcal{S}$-topologies on X^*.

Defininition. (1) If $\mathcal{S} = \mathcal{S}_f$, the collection of all finite subsets of X, then $t_\mathcal{S}$ is the *weak**-*topology* on X^* and is denoted by $w^* = w(X^*, X)$ (or $w(X^*, X)$ and in short by w^*).

(2) [**Sch71, p. 131**] If $\mathcal{S} = \mathcal{S}_{wc}$, the collection of all $w(X, X^*)$-compact subsets of X, then $t_\mathcal{S}$ is called the *Mackey topology* and is denoted by $m^* = m(X^*, X)$.

(3) [**Sch71, p. 140**] If $\mathcal{S} = \mathcal{S}_{wb}$, the collection of all $w(X, X^*)$-bounded subsets of X, then $t_\mathcal{S}$ is called the *strong**-*topology* and is denoted by $b^* = b(X^*, X)$.

Notes. (i) Since every finite set is compact and every compact set is bounded in any TVS, $\mathcal{S}_f \subseteq \mathcal{S}_{wc} \subseteq \mathcal{S}_{wb}$ and so

$$w(X^*, X) \subseteq m(X^*, X) \subseteq b(X^*, X).$$

(ii) If (X, X^*) is a dual pair, then so is (X^*, X). Hence the roles of X and X^* can clearly be interchanged and therefore the topologies $t_\mathcal{S}$ on X^* have their analogues on X by considering $\mathcal{S} \subseteq \mathcal{S}_{w^*b}(X^*)$, the collection of all $w(X^*, X)$-bounded subsets of X^*. In particular, we obtain the topologies

$w(X, X^*)$, $m(X, X^*)$ and $b(X, X^*)$ on X with

$$w(X, X^*) \subseteq m(X, X^*) \subseteq b(X, X^*).$$

Recall that if X and Y be TVSs, then a subset H of $CL(X, Y)$ is said to be **equicontinuous** if for each neighborhood V of 0 in Y, there exists a neighborhood U of 0 in X such that $T(U) \subseteq V$ for all $T \in H$. In particular, we need to have:

Definition. Let (X, τ) be a TVS. A subset H of X^* is said to be **equicontinuous** if, given any $\varepsilon > 0$, there exists a neighbourhood U of 0 in X such that

$$|f(x)| < \varepsilon \quad \text{for all } x \in U \ \text{ and } \ f \in H.$$

Theorem A.4.8. *Let (X, τ) be a TVS. Then*

(i) *For any τ-neighbouhood U of 0 in X, its polar U^o is an equicontinuous subset of X^*.*

(ii) *[**Hor66, p. 200**] A subset H of X^* is equicontinuous $\Longleftrightarrow$ there exists a neighbourhood U of 0 in X such that $H \subseteq U^o$.*

By using the bipolar theorem, we have

Theorem A.4.9. [Sch71, p. 126] *Let (X, τ) be a LCS. Then $\tau = t_{\mathcal{S}_X}$, wher $\mathcal{S}_X$ is the collection of all equicontinuous subsets of X^*. Hence every locally convex topology is a polar topology.*

Definition. [Sch71, p. 130] Let (X, X^*) be a dual pair. A topology τ on X is called *compatible* (or *consistent*) with the dual pair if τ is locally convex topology and $(X, \tau)^* \equiv X^*$.

Theorem A.4.10. [Sch71, p. 132] *Let (X, X^*) be a dual pair and τ a compatible topology on X. Then a subset M of X is τ-bounded $\Longleftrightarrow M$ is $w(X, X^*)$-bounded.*

Theorem A.4.11. *Let (X, X^*) be a dual pair. Then:*

(a) *$w(X, X^*)$ is compatible with the dual pair and is the smallest compatible topology for X.*

(b) *[**Sch71, p. 131**] The largest compatible topology on X is the Mackey topology $m(X, X^*)$.*

(ii) *[**Sch71, p. 132**] A locally convex topology τ on X is compatible $\Longleftrightarrow w(X, X^*) \subseteq \tau \subseteq m(X, X^*)$.*

Note. [Sch71, p. 143; Hor66, p. 209] In general, the strong*-topology $b(X^*, X)$ on X^* need not be compatible.

Definition. [Sch71, p. 132] A locally convex space (X, τ) is called a *Mackey space* if $\tau = m(X, X^*)$.

Theorem A.4.12. [Sch71, p. 132] *Every barrelled or bornological (hence metrizable) LCS is a Mackey space.*

Theorem A.4.13. [Hor66], p. 200] *Let (X, τ) be a TVS. Then*

(a) *[Hor66, p. 200] A subset H of X^* is equicontinuous iff there exists a τ-neighborhood U of 0 in X such that $H \subseteq U^o$. In particular, for any τneighborhood U of 0 in X, U^o is an equicontinuous subset of X^*.*

(b) *Every equicontinuous subset H of X^* is $w(X^*, X)$-bounded.*

Theorem A.4.14. (Bourbaki-Alaoglu) [**Hor66, p. 201**] *Let (X, τ) be a TVS. Then any τ-equicontinuous subset H of X^* is relatively $w(X^*, X)$-compact. In particular, for any τ neighborhood U of 0 in X, U^o is $w(X^*, X)$-compact.*

Corollary A.4.15. (Banach-Alaoglu) *Let X be a normed space. Then the norm closed unit ball*

$$B_1^* = B^*[0, 1] = \{f \in X^* : \|f\| \leq 1\}$$

of X^* is $w(X^*, X)$-compact.

Note. [**Ko69, p. 282**] A norm closed unit ball of a normed space X need not be $w(X, X^*)$-compact in X; take, for example, $X = \ell^1$.

Theorem A.4.16. [**Sim63, p. 234**] *Let $(X, \| \cdot \|)$ is a Banach space, and let $S^* = (B_1^*, w^*)$, a compact Hausdorff space. For any $x \in X$, define a continuous map $\widehat{x} : S^* \to \mathbb{K}$ by*

$$\widehat{x}(f) = f(x), \ f \in S^*.$$

Then the map $\pi : x \longrightarrow \widehat{x}$ is an isometric isomorphism of X onto a closed vector subspace of the Banach space $C(S^*)$.

Remarks. (1) By above theorem, any general Banach space $(X, \|\cdot\|)$ can be considered as a closed vector subspace of the function space $(C(X), \|\cdot\|_\infty)$, where X is a compact Hausdorff space.

(2) [**Jam74, p. 273**] It is fair to ask whether $C[0, 1]$ is typical of many other spaces $C(X)$. In this regard, we mention a famous result of A.A. Milyutin (1966) which states that: *if X is a uncountable compact metric space, then $C(X)$ is isomorphic to $C(I)$.*

Reflexive and Semi-reflexive Spaces

Definition: Let (X, τ) be an LCS.

(i) If $\pi : X \longrightarrow X^{**}$ is onto (i.e. if $\widehat{X} = X^{**}$), then X is called *semireflexive.*

(ii) If $\pi : X \longrightarrow X^{**}$ is a homeomorphism, then X is called *reflexive,* where X^{**} is endowed with the strong*-topology $b^* = b(X^{**}, X^*)$.

Example. For $1 < p < \infty$, the Banach spaces ℓ_p and L_p are reflexive.

Note. If X is a normed space, then $\pi : X \longrightarrow \widehat{X}$ is a homeomorphism. Hence a normed space X is reflexive $\Longleftrightarrow X$ is semi-reflexive.

Theorem A.4.17. *Let (X, τ) be an LCS. Then:*

(a) [**KN76, p. 194**] $\pi : X \longrightarrow X^{**}$ *is one-one $\Longleftrightarrow X$ is Hausdorff.*

(b) [**Sch71, p. 149**] $\pi : X \longrightarrow X^{**}$ *is onto $\Longleftrightarrow X$ is semi-reflexive $\Longleftrightarrow$ every bounded subset of X is relatively $w(X, X^*)$-compact.*

(c) [**KN76, p. 194; Hor66, p. 229**] $\pi : X \longrightarrow X^{**}$ *is a homeomorphism $\Longleftrightarrow X$ is reflexive $\Longleftrightarrow \pi : X \longrightarrow X^{**}$ is onto and continuous.*

Theorem A.4.18. [**KN76, Edw65**] *Let (X, τ_X) and (Y, τ_Y) be LCSs and $T : X \longrightarrow Y$ a linear mapping. Then:*

(a) *If T is continuous (i.e. $\tau_X - \tau_Y$ continuous), then T is also weakly continuous (i.e. w-w continuous), but the converse need not hold* [**KN76, p. 204; Edw65, p. 515, 517**].

(b) *T has an adjoint T^* (i.e T^* exists) $\iff T^*(Y^*) \subseteq X^* \iff T$ is weakly continuous. In this case, T^* is unique. (Hence, if $T : X \longrightarrow Y$ is continuous, then T is weakly continuous and so $T^* : Y^* \longrightarrow X^*$ exists.)* [**KN76, p. 204; Edw65, p. 515**].

(c) *If T has an adjoint T^*, then $T^* : Y^* \longrightarrow X^*$ is $w^* - w^*$ continuous* [**KN76, p. 204; Edw65, p. 523**].

Note. As in part (a) above, if T is $\tau_X - \tau_Y$ continuous, then T is also w-w continuous, but the converse need not hold [**Edw65, p. 517**]. Further, T need not be $w - \tau_Y$ continuous (e.g. the identity map $I : X \longrightarrow X$ is $\tau_X - \tau_X$ continuous and also $w - w$ continuous but not $w - \tau_X$ continuous if $w \neq \tau_X$ (i.e. if $w \subsetneq \tau_X$)).

Projective and Inductive Topologies

Projective Topologies

Definition. Let X be a vector space, $\{(Y_\alpha, \tau_\alpha) : \alpha \in I\}$ a family of TVS, and, for each $\alpha \in I$, $f_\alpha : X \to Y_\alpha$ a linear mapping. The *projective topology* τ_p on X w.r.t. the family $\{(Y_\alpha, \tau_\alpha, f_\alpha) : \alpha \in I\}$ is the coarsest topology on X w.r.t which each f_α is continuous.

Remarks (1) τ_p is the upper bound (in the lattice of topologies on X) of the topologies $f_\alpha^{-1}(\tau_\alpha)$, $\alpha \in I$.

(2) If $x \in X$ and $f_\alpha(x) = x_\alpha \in Y_\alpha$, a τ_p-neighborhood of x is of the form $\bigcap\limits_{i=1}^{n} f_{\alpha_i}^{-1}(U_{\alpha_i})$, where U_{α_i} is any neighborhood of x_α.

(3) If $\mathcal{U}_\alpha$ is a base of τ_α-neighborhood of 0 in Y_α, then a base of τ_α-neighborhoods of 0 in X is given by

$$\mathcal{U}_p = \{\bigcap_{i=1}^{n} f_{\alpha_i}^{-1}(V_{\alpha_i}) = \alpha_1, ..., \alpha_n \in I, \ V_{\alpha_i} \in \mathcal{U}_{\alpha_i}, \ n \geq 1\}.$$

(4) If all τ_α are locally convex, then so is τ_p. In this case, if each τ_α is determined by a family $\mathcal{P}_\alpha$ of seminorms, then τ_p is defined by the family of seminorms

$$\mathcal{P}_{\tau_p} = \{\max_{1 \leq i \leq n} p_{\alpha_i} \circ f_{\alpha_i} : \alpha_1, ..., \alpha_n \in I, \ p_{\alpha_i} \in \mathcal{P}_{\alpha_i}, \ n \geq 1\}.$$

(5) If all τ_α are Hausdorff, then τ_p is Hausdorff if

$$\bigcap_{\alpha \in I} f_\alpha^{-1}(0) = \{0\} \quad (\text{i.e.} \bigcap_{\alpha \in I} \ker f_\alpha = \{0\}).$$

(6) If I is countable and each τ_α is metrizable, then τ_p is semimetrizable.

Theorem A.4.19. Let X be a topological and (X, τ_p) the TVS endowed with the projective topology τ_p defined w.r.t. the family $\{(Y_\alpha, \tau_\alpha, f_\alpha) : \alpha \in I\}$, and let $\varphi : X \to X$ be a mapping. Then φ is continuous iff, for each $\alpha \in I$, $f_\alpha \circ \varphi : X \to (Y_\alpha, \tau_\alpha)$ is continuous.

Proof. [Sch71, p.51] ($\Longrightarrow$) If φ is continuous, then, for each $\alpha \in I$, $f_\alpha \circ \varphi$ (being the composition of two continuous mappings) is continuous.

($\Longleftarrow$) Suppose $f_\alpha \circ \varphi : X \to (Y_\alpha, \tau_\alpha)$ is continuous for each $\alpha \in I$. Let G be an open set in (Y, τ_p). Then we can write

$$G = \bigcap_{i=1}^{n} f_{\alpha_i}^{-1}(G_{\alpha_i}),$$

where each G_{α_i} is open in (Y_α, τ_α). Now

$$\varphi^{-1}(G) = \bigcap_{i=1}^{n} \varphi^{-1}(f_{\alpha_i}^{-1}(G_{\alpha_i})) = \bigcap_{i=1}^{n} (f_{\alpha_i} \circ \varphi)^{-1}(G_{\alpha_i}).$$

Since each $f_\alpha \circ \varphi$ are continuous, it follows that each $(f_\alpha \circ \varphi)^{-1}(G_{\alpha_i})$ is open in X. Thus $\varphi^{-1}(G)$ is open in X. Hence φ is continuous.

Theorem A.4.20. [Won92, p.153], [Sch71, p.53] Every LCS X is isomorphic to a projective limit of a family of Banach spaces.

Proof. Let (X, τ) be a LCS and $\{p_\alpha : \alpha \in I\}$ a family of continuous seminorms on X which determines τ and satisfies $p_\alpha \leq p_\beta$ whenever $\alpha \leq \beta$. For any $\alpha \in I$, let

$$Y_\alpha = X/p_\alpha^{-1}(0), \ \|\cdot\|_\alpha = p_\alpha(\cdot) \quad \text{(the quotient norm on } Y_\alpha\text{)}.$$

$\widehat{Y}_\alpha$ or $(\widehat{Y}_\alpha, \|\widehat{\cdot}\|_\alpha)$ the completion of $(Y_\alpha, \|\cdot\|_\alpha)$ and $\phi_\alpha : X \to \widehat{Y}_\alpha = (\widehat{Y/p_\alpha^{-1}(0)})$ the quotient map. For $\alpha \leq \beta$, let $g_{\alpha\beta} : Y_\beta \to Y_\alpha$ be the canonical map. Then

$$\phi_\alpha = g_{\alpha\beta} \circ \phi_\beta \text{ whenever } \alpha \leq \beta, \tag{1}$$

and τ_p the projective topology w.r.t. $\{(Y_\alpha, \|\cdot\|_\alpha, \phi_\alpha) : \alpha \in I\}$. Define $T : X \to \prod_{\alpha \in I} Y_\alpha$ by

$$Tx = \{\phi_\alpha(x) : \alpha \in I\}.$$

Then it follows from (1) that T is a topological isomorphism from (X, τ) onto a subspace of the projective limit $\varprojlim g_{\alpha\beta}(\widehat{Y}_\beta, \|\widehat{\cdot}\|_\beta)$.

Inductive Topologies [Won92, p.153; Sch71, p.54]

Let Y be a vector space, $\{(X_\alpha, \tau_\alpha) : \alpha \in I\}$ a family of LCS, and for each $\alpha \in I$, $g_\alpha : X_\alpha \to Y$ a linear map. The **inductive topology** τ_i on Y w.r.t. $\{(X_\alpha, \tau_\alpha, g_\alpha) : \alpha \in I\}$ is the finest (strongest) locally convex topology on Y for which each g_α is continuous.

Note. (1) The class $\mathcal{J}$ of locally convex topologies on Y for which all $g_\alpha : X_\alpha \to Y$ are continuous is non-empty (since it contains the indiscrete topology). Now the upper bound τ of $\mathcal{J}$ (i.e. $\tau = \sup \mathcal{J}$) (in the lattice of topologies on Y) is clearly a locally convex topology on Y for which all g_α are continuous. Further, as an upper bound, τ is also a projective topology on Y w.r.t. the family $\{(Y_\lambda, \tau_\lambda, X) : \lambda \in J\}$, where J is an indexing set for

$\mathcal{J}$, $Y_\lambda = Y$ for each $\lambda \in J$ and $e = Y \to Y_\lambda$ is the identity mapping and τ_λ are topologies on the same space $Y_\lambda = Y$.

(2) τ_i need not be Hausdorff, even if all τ_α are

(3) A base for τ_i-neighborhoods of 0 is given by the family $\mathcal{U}$ of all absorbing, convex and balanced subsets of Y such that, for each $\alpha \in I$ and $U \in \mathcal{U}$, $g_\alpha^{-1}(U)$ is a τ_α-neighborhood of 0 in X_α.

If, in particular, $X =$ the linear hull of $\bigcup\limits_{\alpha \in I} g_\alpha(X_\alpha)$, then we may choose $\mathcal{U} = \{U\}$, where $U = -_\alpha g_\alpha(U_\alpha)$. the convex balanced hull of the union $\bigcup\limits_{\alpha \in I} U_\alpha$ i.e.

$$U = \{\Sigma\, t_\alpha g_\alpha(x_\alpha) : \Sigma |t_\alpha| \leq 1,\ x_\alpha \in U_\alpha\},$$

where U_α is any member of a base of τ_α-neighborhoods of 0 in X_α.

Theorem A.4.21. Let Y be an LCS and (Y, τ_i) the space Y with the inductive topology τ_i w.r.t. the family $\{(X_\alpha, \tau_\alpha, g_\alpha) : \kappa \in I\}$. Then linear map $\psi : (Y, \tau_i) \to Y$ is continuous iff each $\psi \circ g_\alpha = (X_\alpha, \tau_\alpha) \to Y$ is continuous.

Proof. [Sch71, p.54] ($\Longrightarrow$) If ψ is continuous, then, for each $\alpha \in I$, $\cdot\, g_\alpha$ (being the composition of two continuous mappings) is continuous.

($\Longleftarrow$) Suppose $\psi \circ g_\alpha$ is continuous for each $\alpha \in I$. Let W be a convex balanced neighborhood of 0 in Y. Then $(\psi \circ g_\alpha)^{-1}(W) = g_\alpha^{-1} \psi^{-1}(W)$ is a neighborhood of 0 in X_α for each $\alpha \in I$, which implies that $\psi^{-1}(W)$ is a neighborhood of 0 in (Y, τ_i). Thus ψ is continuous.

$\{(X_\alpha, \tau_\alpha) : \alpha \in I\}$ of X is called a *strict inductive limit* if $\tau_\beta \mid X_\alpha = \tau_\alpha(\alpha \leq \beta)$.$\}$

A.5 Spaces of Continuous Vector-valued Functions

Uniform, Strict and Compact-open Topologies on $C_b(\mathbf{S}, \mathbf{X})$

Let S be a topological space and X a locally convex space whose topology is generated by a family $cs(X)$ of continuous seminorms. For any $p \in cs(X)$ and $\varepsilon > 0$, we write

$$W_{p,\varepsilon} = \{a \in X : p(a) \leq \varepsilon\},$$

a closed balanced absorbing and balanced set in X. If $\varepsilon = 1$, we shall denote $W_{p,\varepsilon}$ simply by W_p.

Definition: A function $f : S \to X$ is said to **bounded** if $f(S)$ is a bounded subset of X, i.e. for each $p \in cs(X)$, there exists an $r > 0$ such that

$$\|f\|_p = \|p \circ f\| = \sup_{x \in S} p(f(x)) \leq r.$$

$f : S \to X$ is said to **vanish at infinity** if, for each $p \in cs(X)$ and $\varepsilon > 0$, there exists a compact set $K = K_{p,\varepsilon} \subseteq S$ such that

$$p(f(x)) < \varepsilon \text{ for all } x \in S \backslash K;$$

in particular, a function $\varphi : S \to \mathbb{K}$ is said to **vanish at infinity** if, for each $\varepsilon > 0$, there exists a compact set $K = K_\varepsilon \subseteq S$ such that

$$|f(x)| < \varepsilon \text{ for all } x \in S \backslash K.$$

The **support of** $f : S \to X$ is defined as the closure of the set $\{x \in S : f(x) \neq 0\}$ in S and we write

$$supp(f) = cl - \{x \in S : f(x) \neq 0\}.$$

Let $\overline{F_s}(S, X)$ be the vector space of all bounded X-valued functions on S and $C(S, X)$ be the vector space of all continuous X-valued functions on S, and let $C_b(S, X)$ (resp. $C_0(S, X)$, $C_{00}(S, X)$) the subspace of $C(S, X)$ consisting of those functions which are bounded (resp. vanish at infinity, have compact support}. When $X = \mathbb{K}$ ($= \mathbb{R}$ or $\mathbb{C}$), then these spaces will be denoted by $B(S), C(S), C_b(S), C_0(S)$ and $C_{00}(S)$. Let $B_0(S)$ denote the subset of $B(S)$ consisting of those functions which vanish at infinity. We shall denote by $B(S) \otimes X$ the vector space spanned by the set of all functions of the form $\varphi \otimes a$, where $\varphi \in B(S)$, $a \in X$, $(\varphi \otimes a)(x) = \varphi(x)a$ $(x \in S)$. $F(S, X) = X^S$ denotes the vector space of all functions $f : S \to X$.

Clearly, $C_b(S, X) = C(S, X) \cap B(S, X)$, $C_0(S, X) \subseteq C_b(S, X) \subseteq C(S, X)$, with $C_0(S, X) = C_b(S, X) = C(S, X)$ if S is compact. Further, these function spaces are vector spaces over $\mathbb{K}$ under addition $f + g$ and scalar multiplication λf defined pointwise as:

$$(f + g)(x) = f(x) + g(x), \quad (\lambda f)(x) = \lambda f(x), \quad x \in S.$$

We now describe a general method of defining linear topologies on $C_b(S, X)$.

Definition: Let S be a topological space and X a locally convex space whose topology is generated by a family $cs(X)$ of continuous seminorms. For any subset $\mathcal{B}$ of $B(S)$, we define the $\mathcal{B}$-*topology* on $C_b(S, X)$ is defined as the locally convex topology generated by a family of seminorms $\{||\cdot||_{p,\varphi} : p \in cs(X), \varphi \in \mathcal{B}\}$, where

$$||f||_{p,\varphi} = \sup_{x \in S} p(\varphi(x).(f(x))), \quad f \in C_b(S, X).$$

Equivalently, the $\mathcal{B}$-topology $C_b(S, X)$ is the locally convex topology having a subbase of convex neighbourhoods of 0 consisting of all sets of the form $\{N(\varphi, W_{p,\varepsilon}) : p \in cs(X), \varphi \in \mathcal{B} \text{ and } \varepsilon > 0\}$, where

$$
\begin{aligned}
N(\varphi, W_{p,\varepsilon}) &= \{f \in C_b(S, X) : (\varphi f)(S) \subseteq W_{p,\varepsilon}\} \\
&= \{f \in C_b(S, X) : ||f||_{p,\varphi} \leq \varepsilon\}.
\end{aligned}
$$

For any $A \subseteq S$, we shall write

$$
\begin{aligned}
N(A, W_{p,\varepsilon}) \quad &: \quad = N(\chi_A, W_{p,\varepsilon}) = \{f \in C_b(S, X) : f(A) \subseteq W_{p,\varepsilon}\} \\
&= \{f \in C_b(S, X) : ||f||_{p,A} \leq \varepsilon\},
\end{aligned}
$$

Now, we can define the topologies p, k, β, σ, u more directly as follows.

Definition: Let S be a topological space and X a locally convex space whose topology is generated by a family $cs(X)$ of continuous seminorms.

(1) The **uniform topology** u on $B(S, X)$ or $C_b(S, X)$ is defined as the locally convex topology generated by a family of seminorms $\{||\cdot||_p : p \in P\}$, where

$$||f||_p = ||p \circ f|| = \sup_{x \in S} p(f(x)), \quad f \in B(S, X), \text{ or } C_b(S, X).$$

(2) The **strict topology** β [Buck58] on $C_b(S, X)$ is defined as the locally convex topology generated by a family of seminorms $\{||\cdot||_{p,\varphi} : p \in cs(X), \varphi \in B_0(S)\}$, where

$$||f||_{p,\varphi} = \sup_{x \in S} p(\varphi(x).(f(x)), \quad f \in C_b(S, X).$$

(3) The **compact-open topology** k on $C(S, X)$ is defined as the locally convex topology generated by a family of seminorms $\{||\cdot||_{p,K} : p \in cs(X), K \subseteq S\}$, where

$$||f||_{p,K} = ||p \circ f||_K = \sup_{x \in K} p(f(x)), \quad f \in C(S, X),$$

and K varies over all compact subsets of S.

(4) The **pointwise topology** p on $C(S, X)$ is defined as the locally convex topology generated by a family of seminorms $\{||\cdot||_{p,F} : p \in cs(X), F \subseteq S\}$, where

$$||f||_{p,F} = ||p \circ f||_F = \sup_{x \in F} p(f(x)), \quad f \in C(S, X),$$

and F varies over all finite subsets of S.

Lemma A.5.1. [KR91] *Suppose that $A, B \subseteq S$ and that $p, q \in cs(X)$ are such that $N(A, W_q) \subseteq N(B, W_p)$. Then:*

(i) *If $W_p \neq X$, then $B \subseteq \overline{A}$.*

(ii) *If B is non-empty, then $W_q \subseteq W_p$.*

Proof. (i) Suppose $B \not\subseteq \overline{A}$, and let $x \in B \backslash \overline{A}$. Since S is completely regular, there exists a φ in $C_b(S)$ such that $\varphi(x) = 1$ and $\varphi(\overline{A}) = \{0\}$. Then clearly $supp \, (\varphi) \cap \overline{A} = \varnothing$. Choose $a \in X \backslash W_p$ and let $f = \varphi \otimes a$. Then $f(A) = \{0\} \subseteq W_q$, and so $f \in N(A, W_q)$; but $f(x) = a \notin W_p$ and so $f \notin N(B, W_p)$. This contradiction implies that $B \subseteq \overline{A}$.

(ii) Suppose $W_q \subsetneq W_p$. Choose $b \in W_q \backslash W_p$. and let $g = \chi_X \otimes b$. If $A = \varnothing$, then $g(A) = \varnothing \subseteq W_q$; if $A \neq \varnothing$, then $g(A) = \{b\} \subseteq W_q$. Hence $g \in N(A, W_q)$. Since $B \neq \varnothing$, $g(B) = \{b\} \not\subseteq W_p$ and so $g \notin N(B, W_p)$, a contradiction. $\square$

Theorem A.5.2. [Buc58, Kh79] (a) $p \leq k \leq \beta \leq \sigma \leq u$.

(b) *If S is completely regular, then:*

(i) $u = \beta$ *iff S is compact.*

(ii) $\beta = k$ iff every σ-compact subset of S is relatively compact.

(c) u and β have the same bounded sets in $C_b(S, X)$.

(d) $\beta = k$ on u-bounded subsets of $C_b(S, X)$.

(e) A sequence $\{f_n\}$ in $C_b(S, X)$ is β-convergent iff it is u-bounded and k-convergent.

(f) If $X = (X, d)$ is metrizable, then $(B(S, X), u)$ and $(C_b(S, X), u)$ are also metrizable with respect to the metric given by:

$$\rho(f, g) = \sup_{x \in S} d(f(x), f(x)), \ f, g \in B(S, X) \ \text{ or } \ C_b(S, X).$$

Remark. If S is not locally compact, then $C_0(S, X)$ may be the trivial vector space $\{0\}$.

Counter-example. Let $S = \mathbb{Q}$, the space of rationals and $X = \mathbb{R}$. Then $C_0(\mathbb{Q}, \mathbb{R}) = \{0\}$, as follows. Suppose $\varphi(\neq 0) \in C_0(\mathbb{Q}, \mathbb{R})$. Choose $y \in Q$ with $\varphi(y) \neq 0$. We may suppose $\varphi(y) > 0$. Choose $a, b > 0$ such that $a < \varphi(y) < b$. Then the set

$$U = \{x \in \mathbb{Q} : a < \varphi(x) < b\} = \varphi^{-1}(a, b)$$

is non-empty (since $y \in U$) and open (since φ is continuous). Since $\varphi \in C_0(\mathbb{Q}, \mathbb{R})$, the set

$$V = \{x \in \mathbb{Q} : |\varphi(x)| > a\} = \{x \in \mathbb{Q} : \varphi(x) > a \text{ or } \varphi(x) < -a\}$$

is relatively compact. Now, for any $x \in U$, $\varphi(x) > a$ and so $x \in V$; that is, $U \subseteq V$. Then it follows that $\overline{U}$ is compact. Hence every point in U has a compact neighborhood. Upon translating U along $\mathbb{Q}$, we arrive at the contradiction that $\mathbb{Q}$ is locally compact. $\square$

Some Approximation theorems for C(S,X) and C_b(S,X)

In this section, we present two lemmas on "Partition of Unity". These play a fundamental role in the proofs of the Stone-Weierstrass type theorems for vector-valued functions.

Lemma A.5.3. (Partition of Unity) [**Rud91, p. 40**] *Let S be a locally compact Hausdorff space and K a compact subset of S Then, for any open cover $\{U_1, ..., U_n\}$ of K, there exist $\phi_1, ..., \phi_n \in C_b(S)$ such that*

$$0 \le \phi_i \le 1, \ \text{supp } \phi_i \subseteq U_i, \ \sum_{i=1}^{n} \phi_i = 1 \text{ on } K \text{ and } \sum_{i=1}^{n} \phi_i \le 1 \text{ on } S.$$

Proof. Let $x \in K$. Since $K \subseteq \bigcup_{i=1}^{n} U_i$, $x \in U_i$ for some $1 \le i \le n$. Since S be a locally compact Hausdorff space, there exists a neighbourhood V_x of x in S such that cl-V_x is compact and cl-$V_x \subseteq U_i$. Clearly, $\{V_x : x \in K\}$ is an open cover of compact K, and so it has a finite subcover $\{V_{x_j} : 1 \le j \le m\}$ (say) for K. For each $1 \le i \le n$, let

$$K_i = \bigcup \{\overline{V_{x_j}} : \overline{V_{x_j}} \subseteq U_i, 1 \le j \le m\}.$$

Clearly, $K \subseteq \bigcup_{i=1}^{n} K_i$. Further, each K_i is compact and $K_i \subseteq U_i$. Since S is locally compact Hausdorff, it is normal. Since K_i and $S \backslash U_i$ are closed disjoint sets in S, by the Urysohn's lemma, there exist functions $\psi_1, ..., \psi_n \in C_b(S)$ such that

$$0 \le \psi_i \le 1, \quad \psi_i = 1 \text{ on } K_i \text{ and } \psi_i = 0 \text{ on } S \backslash U_i, \quad (1 \le i \le 1).$$

Define

$$
\begin{aligned}
\phi_1 &= \psi_1 \\
\phi_2 &= (1 - \psi_1)\psi_2 \\
&\quad \cdots\cdots\cdots \\
\phi_n &= (1 - \psi_1)(1 - \psi_2)....(1 - \psi_{n-1})\psi_n \quad (n \ge 2)
\end{aligned}
$$

Clearly, $0 \le \phi_i \le 1$ and $\phi_i = 0$ on $S \backslash U_i$ and so supp $\phi_i \subseteq U_i$. By induction, we can write

$$
\begin{aligned}
\sum_{i=1}^{n} \phi_i &= \psi_1 + (1 - \psi_1)\psi_2 + ... + (1 - \psi_1)(1 - \psi_2)....(1 - \psi_{n-1})\psi_n \\
&= 1 - (1 - \psi_1)(1 - \psi_2)....(1 - \psi_n) = 1 - \prod_{i=1}^{n}(1 - \psi_i)
\end{aligned}
$$

If $x \in S$. Then clearly

$$\sum_{i=1}^{n} \phi_i(x) = 1 - \prod_{i=1}^{r} [1 - \psi_i(x)] \le 1 - 0 = 1.$$

Let $x \in K$. Since $K \subseteq \bigcup_{i=1}^{n} K_i$, $x \in K_j$ for some $1 \le j \le n$, and so $\psi_j(x) = 1$. Hence

$$\sum_{i=1}^{n} \phi_i(x) = 1 - (1 - \psi_j(x)) \prod_{i \ne j} [1 - \psi_i(x)] = 1 - 0 = 1.$$

Lemma A.5.4. (Partition of Unity) *Let S be a completely regular space, and let K be a compact subset of S Then, given any open cover $\{U_1, ..., U_n\}$ of K, there exist $\varphi_1, ..., \varphi_n \in C_b(S)$ such that*

$$0 \le \varphi_i \le 1, \; \varphi_i = 0 \text{ outside } U_i, \; \sum_{i=1}^{n} \varphi_i = 1 \text{ on } K, \text{ and } \sum_{i=1}^{n} \varphi_i \le 1 \text{ on } S.$$

Proof. By $(*)$, for each $t \in K$, there is an $f_t \in C_b(S)$ such that $f_t(t) \ne 0$. We may assume that $f_t \ge 0$. [Indeed, let $\gamma : \mathbb{K} \longrightarrow \mathbb{K}$ be defined by

$$
\gamma(x) = \begin{cases} 0 & if & x = 0, \\ |x|^2/x & if & 0 < |x| < 1, \\ |x|/x & if & |x| \ge 1. \end{cases}
$$

Clearly, γ is bounded, continuous and satisfies $\gamma(x) \neq 0$ for $x \neq 0$ and $\gamma(x)x \geq 0$ for $x \in \mathbb{K}$. Set $\phi = \gamma \circ f_t \in C_b(S)$; then $\phi(x) \neq 0$ and $\phi f_t \geq 0$. We replace f_t by $\phi f_t \in C_b(S)$ to achieve $f_t \geq 0$]. We may also assume that $f_t = 0$ outside some U_i. [In fact, choose some U_i containing t and $\psi \in C_b(S)$ such that $0 \leq \psi \leq 1$, $\psi(t) = 1$ and $\psi = 0$ outside U_i. Therefore, it would be sufficient to replace f_t by $\psi f_t \in C_b(S)$ to assure that $f_t = 0$ outside some U_i.]

By compactness of K, there exists a finite collection $\{f_1, ..., f_m\} \subseteq \{f_t : t \in K\} \subseteq C_b(S)$ such that $f_j \geq 0$ on S and $f_j = 0$ outside some U_j $(j = 1, ..., m)$ and $\sum_{j=1}^{m} f_j(x) > 0$ for all $x \in K$. For each $i = 1, ..., n$, define $J_i = \{j : f_j = 0 \text{ outside } U_i, 1 \leq j \leq m\}$, and let $g_i = \sum_{j \in J_i} f_j$. Then $g_i \in C_b(S)$,

$$g_i \geq 0, \ g_i = 0 \ \text{ outside } \ U_i \ \ (1 \leq i \leq n) \ \text{ and } \ \sum_{i=1}^{n} g_i(x) > 0 \ \text{ for all } \ x \in K.$$

By compactness of K, there is an $h \in C_b(S)$ such that

$$0 \leq h \leq 1 / \sum_{i=1}^{n} g_i \ \text{ on } \ S \ \text{ and that } \ h = 1 / \sum_{i=1}^{n} g_i \ \text{ on } \ K.$$

Setting $\varphi_i = h g_i$, we clearly have

$$0 \leq \varphi_i \leq 1, \ \varphi_i = 0 \ \text{ outside } \ U_i, \ \sum_{i=1}^{n} \varphi_i = 1 \ \text{ on } \ K \ \text{ and } \ \sum_{i=1}^{n} \varphi_i \leq 1 \ \text{ on } \ S. \ \square$$

Theorem A.5.5. *Let S be a locally compact Hausdorff space and X an LCS (in particular, a normed space). Then $C_0(S) \otimes X$ is u-dense in $C_0(S, X)$.*

Proof. Let $f \in C_0(S, X)$, and let p be any continuous seminorm on X and $\varepsilon > 0$. Since $f \in C_0(S, X)$, there exists a compact set $K \subseteq S$ such that

$$p(f(x)) < \varepsilon \text{ for all } x \notin K. \tag{1}$$

For each $x \in X$, let

$$N(x) = f^{-1}(f(x) + W_{p,\varepsilon}) = \{y \in X : p(f(y) - f(x)) < \varepsilon\}. \tag{2}$$

The sets in $\{N(x) : x \in K\}$ form an open cover of K, and so there exists a finite open cover, $\{N(x_i) : i = 1, ..., n\}$, (say). By partition of unity Lemma A.2.3, there exist $\phi_1, ..., \phi_n \in C_0(S)$ such that $0 \leq \phi_1 \leq 1$, $\phi_i = 0$ outside of $N(x_i)$,

$$\sum_{i=1}^{n} \phi_i(x) = 1 \text{ for } x \in K, \text{ and } \sum_{i=1}^{n} \phi_i(x) \leq 1 \text{ for } x \in S. \tag{3}$$

Choose $y_i \in N_i$ $(1 \leq i \leq n)$. Define an X-valued function $g : S \to X$ by

$$g = \sum_{i=1}^{n} \phi_i \otimes f(y_i).$$

Clearly, $g \in C_0(S) \otimes X$. Let y be any point in S, and let $I_y = \{i : y \in N(x_i)\}$. Note that $p(f(y) - f(y_i)) \leq \varepsilon$ if $i \in I_y$ and $\varphi_i(y) = 0$ if $i \notin I_y$. Consequently, if $y \in K$, then by (3) $\sum_{i=1}^{n} \phi_i(y) = 1$ and so

$$
\begin{aligned}
p(g(y) - f(y)) &= p\left(\sum_{i=1}^{n} \phi_i(y).f(y_i) - f(y) \right) \\
&= p\left(\sum_{i=1}^{n} \phi_i(y).f(y_i)) - (\sum_{i=1}^{n} \phi_i(y))f(y)) \right) \\
&= p\left(\sum_{i=1}^{n} \phi_i(y).(f(y_i) - f(y)) \right) \\
&\leq \sum_{i \in I_y} \phi_i(y).p\left((f(y_i) - f(y)) \right) \leq \sum_{i \in I_y} \phi_i(y)\varepsilon \leq \varepsilon.
\end{aligned}
$$

If $y \notin K$, then by (3) $\sum_{i=1}^{n} \phi_i(y) \leq 1$, so using (1) and (2),

$$
\begin{aligned}
p(g(y) - f(y)) &= p\left(\sum_{i=1}^{n} \phi_i(y).(g_i(y_i) - f(y)) + \{\sum_{i=1}^{n} \phi_i - 1\}.f(y) \right) \\
&\leq \sum_{i=1}^{n} \phi_i(y).p(g_i(y_i) - f(y)) + |\sum_{i=1}^{n} \phi_i - 1|.p(f(y)) \\
&\leq \sum_{i=1}^{n} \phi_i(y).\varepsilon + |\sum_{i=1}^{n} \phi_i - 1|\varepsilon \leq \varepsilon + \varepsilon = 2\varepsilon.
\end{aligned}
$$

Since $y \in S$ is arbitrary, $\|g - f\|_p \leq 2\varepsilon$. Since $g \in C_0(S) \otimes X$, we have $f \in \overline{C_0(S) \otimes X}$. This shows that A is u-dense in $C_0(S, X)$. $\square$

Theorem A.5.6. *Let S be a completely regular Hausdorff space and X an LCS (in particular, a normed space). Then $C(S) \otimes X$ is k-dense in $C(S, X)$.*

Proof. Let $f \in C(S, X)$, and let $p \in cs(X)$, $K \subseteq S$ a compact set and $\varepsilon > 0$. For each $x \in X$, let

$$
N(x) = f^{-1}(f(x) + W_{p,\varepsilon}) = \{y \in X : p(f(y) - f(x)) < \varepsilon\}. \qquad (1)
$$

The sets in $\{N(x) : x \in K\}$ form an open cover of K, and so there exists a finite open cover, $\{N(x_i) : i = 1, .., n\}$, (say). By partition of unity Lemma A.2.4, there exist $\varphi_1, ..., \varphi_n \in C(S)$ such that $0 \leq \varphi_1 \leq 1$, $\varphi_i = 0$ outside of $N(x_i)$,

$$
\sum_{i=1}^{n} \varphi_i(x) = 1 \text{ for } x \in K, \text{ and } \sum_{i=1}^{n} \varphi_i(x) \leq 1 \text{ for } x \in S. \qquad (2)
$$

Choose $y_i \in N(x_i)$ $(1 \leq i \leq n)$. Define an X-valued function $g : S \to X$ by

$$
g = \sum_{i=1}^{n} \varphi_i \otimes f(y_i).
$$

Clearly, $g \in C(S) \otimes X$. Let y be any point in S, and let $I_y = \{i : y \in N(x_i)\}$. Note that $p(f(y) - f(y_i)) \leq \varepsilon$ if $i \in I_y$ ==and $\varphi_i(y) = 0$ if $i \notin I_y$. Consequently, if $y \in K$, then by (2) $\sum_{i=1}^n \varphi_i(y) = 1$ and so

$$
\begin{aligned}
p(g(y) - f(y)) &= p\left(\sum_{i=1}^n \varphi_i(y)(g_i(y) - f(y))\right) \\
&\leq \sum_{i \in I_y} \varphi_i(y) p(g_i(y) - f(y)) \\
&< \sum_{i \in I_y} \varphi_i(y)\varepsilon \leq \varepsilon.
\end{aligned}
$$

Thus $||g - f||_{p,K} \leq \varepsilon$, and so f belongs to the k-closure of $C(S) \otimes X$; that is, $C(S) \otimes X$ is k-dense in $C(S, X)$, as required. $\square$

Let $B_0(S)$ denote the set of all bounded functions $\varphi : S \to \mathbb{K}$ which vanish at infinity. The *strict topology* β on $C_b(S, X)$ is the locally convex topology generated by a family of seminorms $\{|| \cdot ||_{\varphi,p} : p \in cs(X), \varphi \in B_0(S)\}$, where

$$
||f||_{\varphi,p} = \sup_{x \in S} p(\varphi(x).(f(x))), \quad f \in C_b(S, X).
$$

Theorem A.5.7. *Let S be a completely regular Hausdorff space and X an LCS (in particular, a normed space). Then $C_b(S) \otimes X$ is β-dense in $C_b(S, X)$.*

Proof. Let $f \in C_b(S, X)$, and let p be any continuous seminorm on X, $\varphi \in B_0(S)$, $0 \leq \varphi \leq 1$ and $\varepsilon > 0$. There exists a compact set $K \subseteq S$ such that

$$
p(\varphi(x)f(x)) < \varepsilon \text{ for all } x \notin K. \tag{1}
$$

For each $x \in X$, let

$$
N(x) = f^{-1}(f(x) + W_{p,\varepsilon}) = \{y \in X : p(f(y) - f(x)) < \varepsilon\}. \tag{2}
$$

The sets in $\{N(x) : x \in K\}$ form an open cover of K, and so there exists a finite open cover, $\{N(x_i) : i = 1, ..., n\}$, (say). By **Partition of Unity Lemma A.2.4**, there exist $\varphi_1, ..., \varphi_n \in C_0(S)$ such that $0 \leq \varphi_1 \leq 1$, $\varphi_i = 0$ outside of $N(x_i)$,

$$
\sum_{i=1}^n \varphi_i(x) = 1 \text{ for } x \in K, \text{ and } \sum_{i=1}^n \varphi_i(x) \leq 1 \text{ for } x \in S. \tag{3}
$$

Choose $y_i \in N_i$ $(1 \leq i \leq n)$. Define an X-valued function $g : S \to X$ by

$$
g = \sum_{i=1}^n \varphi_i \otimes f(y_i).
$$

Clearly, $g \in C_b(S) \otimes X$. Let y be any point in S, and let $I_y = \{i : y \in N(x_i)\}$. Note that $p(f(y) - f(y_i)) \leq \varepsilon$ if $i \in I_y$ and $\varphi_i(y) = 0$ if $i \notin I_y$.

Consequently, if $y \in K$, then by **(3)** $\sum_{i=1}^{n} \varphi_i(y) = 1$ and so

$$
\begin{aligned}
p(\varphi(y)(g(y) - f(y))) \;&=\; \varphi(y)p\left(\sum_{i=1}^{n} \varphi_i(y)g_{x_{j_i}}(y) - f(y) \right) \\
&=\; \varphi(y)p\left(\sum_{i=1}^{n} \varphi_i(y)g_{x_{j_i}}(y) - (\sum_{i=1}^{n}\varphi_i(y))f(y) \right) \\
&=\; \varphi(y)p\left(\sum_{i=1}^{n} \varphi_i(y)(g_{x_{j_i}}(y) - f(y)) \right) \\
&\leq\; \varphi(y) \sum_{i \in I_y} \varphi_i(y)p(g_{x_{j_i}}(y) - f(y)) \\
&<\; \sum_{i \in I_y} \varphi_i(y).\varepsilon \leq \varepsilon.
\end{aligned}
$$

If $y \notin K$, we have

$$
\begin{aligned}
p(\varphi(y)(g(y) - f(y))) \;&=\; p(\varphi(y) \sum_{i=1}^{n} \varphi_i(y)(g_{x_{j_i}}(y) - f(y)) + \sum_{i=1}^{n} \varphi_i(y)\varphi(y)f(y) - \varphi(y) \\
&=\; p(\varphi(y) \sum_{i=1}^{n} \varphi_i(y)(g_{x_{j_i}}(y) - f(y)) + \{\sum_{i=1}^{n} \varphi_i(y) - 1\}\varphi(y)f(y)) \\
&\leq\; \varphi(y) \sum_{i=1}^{n} \varphi_i(y)p(g_{x_{j_i}}(y) - f(y)) + |\sum_{i=1}^{n} \varphi_i(y) - 1|.p(\varphi(y)f(y)) \\
&<\; \varphi(y).\varepsilon. \sum_{i=1}^{n} \varphi_i(y) + |\sum_{i=1}^{n} \varphi_i(y) - 1|.\varepsilon \leq 2\varepsilon.
\end{aligned}
$$

Thus $\|g - f\|_{p,\varphi} < 2\varepsilon$, and so f belongs to the β-closure of $\mathcal{M}$; that is, $\mathcal{M}$ is β-dense in $C_b(S, X)$, as required. $\square$

To obtain further approximation results, we need some terminology.

Definition: (1) A TVS X is said to have the *approximation property* if the identity map on X can be approximated uniformly on precompact sets by continuous and linear maps of finite rank (i.e. with range contained in finite dimensional subspaces of X).

(2) A TVS X is said to be *admissible* if the identity map on X can be approximated uniformly on compact sets by continuous maps of finite rank.

Every locally convex space and every F-space with a basis (e.g. $\ell_p, 0 < p < 1$) is admissible [**Kl60a**], [**Kl60b**, [**Shu72**]).

Let $C_{pc}(S, X)$ (resp. $C_{rc}(S, X)$) denote the subspace of $C_b(S, X)$ consisting of those functions f such that $f(S)$ is precompact (resp. relatively compact). Note that if S is complete, then $C_{pc}(S, X) = C_{rc}(S, X)$.

Theorem A.5.8. *If S be a completely regular space and E a TVS with a base $\mathcal{W}$ of neifghbourhoods of 0.*

(a) E is admissible iff for all topological spaces S, $C_b(S) \otimes E$ is u-dense in $C_{rc}(S, E)$.

(b) If E has the approximation property, then $C_b(S) \otimes E$ is u-dense in $C_{pc}(S, E)$.

(c) If S is a normal space, then $C_b(S) \otimes E$ is u-dense in $C_{rc}(S, E)$.

(d) If E is an admissible TVS, then

    (i) $C_b(S) \otimes E$ is k-dense in $C(S, E)$.

    (ii) $C_b(S) \otimes E$ is β-dense in $C_b(S, E)$.

Proof. (a) Suppose E is a admissible, and let $f \in C_{rc}(S, E)$ and $W \in \mathcal{W}$ a balanced set. Then there exists a continuous map $u : \overline{f(S)} \to E$ with range contained in a finite dimensional subspace of E such that $u(a) - a \in \overline{f(S)}$. Then $h = u \circ f \in C_b(S) \otimes E$ and $h(x) - f(x) \in W$ for all $x \in S$.

Conversely, let $K \subseteq E$ be a compact set and $W \in \mathcal{W}$. Since, by hypothesis, $C_b(K) \otimes E$ is u-dense in $C_b(K, E)$, there exists some $g = \sum_{i=1}^{n} \varphi_i \otimes a_i$ ($\varphi_i \in C_b(K), a_i \in E$) in $C_b(K) \otimes E$ such that $g(a) - a \in W$ for all $a \in K$. Note that the range of g is contained in the finite dimensional subspace spanned by $\{a_1, ..., a_n\}$, say. Thus E is admissible.

(b) Its proof is similar to the first part of (a).

(c) Since S is normal, its Stone-Cech compactification βS also has finite covering dimension [GJ60]. So, by Corollary 5.1.7, $C_b(\beta S) \otimes E$ is u-dense in $C_b(\beta S, E)$. Note that each function in $C_{rc}(S, E)$ has a continuous extension to all of βS. Hence $C_b(S) \otimes E$ and $C_{rc}(S, E)$ are linearly isomorphic to $C_b(\beta S) \otimes E$ and $C_b(\beta S, E)$, respectively, and so the result follows.

(d) (i) Let $f \in C(S, E)$, and let K be a compact subset of S and $W \in \mathcal{W}$ a balanced set. By hypothesis, there exists a continuous map $u : f(K) \to E$ with range contained in a finite dimensional subspace of E such that $u(f(x)) - f(x) \in W$ for all $x \in K$. We can write

$$u \circ f = \sum_{i=1}^{m} (u_i \circ f) \otimes a_i, \text{ where } u_i \circ f \in C(K) \text{ and } a_i \in E.$$

By the famous Tietze extension Theroem, there exist $h_1, ..., h_m \in C_b(S)$ such that $h_i = u_i \circ f$ on K. Let $g = \sum_{i=1}^{m} h_i \otimes a_i$. Then $g \in C_b(S) \otimes E$ and $g - f \in N(K, W)$.

(d) (ii) Let $f \in C_b(S, E)$, and let $\varphi \in B_0(S)$ and $W \in \mathcal{W}$. Choose an open balanced neighbourhood $V \in \mathcal{W}$ such that $V + V + V \subseteq W$. Choose $r > \|\varphi\|$ with $f(S) \subseteq rV$, and let

$$K = \{x \in X : \varphi(x) \geq 1/r\}.$$

Then $f(K)$ is a compact subset of E and so, by hypothesis, there exists a continuous map $u : f(K) \to E$ with range contained in a finite dimensional subspace of E such that

$$u(f(x)) - f(x) \in (1/r)G \text{ for all } x \in K.$$

We can write

$$u \circ f = \sum_{i=1}^{n} (u_i \circ f)(x)a_i \quad (u_i \circ f \in C(K) \text{ and } a_i \in E).$$

Again there exist $\theta_i (1 \leq i \leq n)$ in $C_b(S)$ such that $\theta_i = u_i \circ f$ on K. Let $h = \sum_{i=1}^{n} \theta_i \otimes a_i$. Then

$$K \subseteq h^{-1}(rV + rV) = F \text{ (say)},$$

which is open in S, and so there exists a $\theta \in C_b(S)$ with $0 \leq \theta \leq 1$, $\theta = 0$ on $S \backslash F$. Let $g = \theta h$. Then $g \in C_b(S) \otimes E$ and $g = h = u \circ f$ on K. Further, $g(S) \subseteq rV + rV$. It is now easily verified that $g - f \in N(\varphi, W)$, as required. $\square$

A.6 Measure Theory

In this section, we give a overview of the famous Riesz representation Theorem regarding characterization of dual spaces of $(C(S), u)$ and $(C_0(S), u)$ as certain spaces $M(S)$, $M_t(S)$ of Borel measures via the integral representation.

First we state the classical Riesz representation Theorem for the dual of space $C[a, b]$. We shall consider the uniform topology u on $C[a, b]$ given by the sup norm:

$$\|f\| = \sup\{|f(t)| : t \in [a.b]\}, \ f \in C[a, b].$$

Definition: Let $g : [a, b] \rightarrow \mathbb{R}$ be a function, and let $\mathcal{P}$ be the collection of all finite partitions of $[a, b]$. Then g is said to be of **bounded variation** if there exists a constant $K > 0$ such that, given any partition $P \in \mathcal{P}$, say

$$P = P_n = \{a = x_0, x_1, ..., x_n = b\},$$

we have

$$V(g, P) = \sum_{i=1}^{n} |g(x_i) - g(x_{i-1})| \leq K.$$

Clearly, every function of bounded variation is bounded, but need not be continuous. Also a bounded or continuous function also need not be of bounded variation. Bounded monotonic functions are of bounded variation. In fact, a function $g : [a, b] \rightarrow \mathbb{R}$ is of bounded variation iff it can be expressed as $g = h - k$, where h, k are monotonic increasing and bounded functions on $[a, b]$ (see Apostal [Ap57])

We shall denote by $BV[a, b]$ the vector space of all bounded variations of $[a, b]$. The total variation of any $g \in BV[a, b]$ is defined by

$$V(g) = \sup\{V(g, P) : P \in \mathcal{P}\}.$$

It is well-known that if $f \in C[a, b]$ and $g \in BV[a, b]$, then the Riemann-Stieltjes integral $\int_{a}^{b} f dg$ exists.

Theorem A.6.1.(Riesz, 1909) For any given $g \in BV[a,b]$, the mapping $L = L_g : C[a,b] \to \mathbb{R}$ given by

$$L(f) = \int_a^b f dg \quad \text{for all } f \in C[a,b], \qquad (*)$$

is a continuous linear functional on $C[a,b]$ (i.e. $L \in (C[a,b],u)^*)$ with $\|L\| = V(g)$. Conversely, if L is a continuous linear functional on $C[a,b]$, then there exists a unique $g \in BV[a,b]$ such that L can be expressed as in $(*)$ with $\|L\| = V(g)$.

In this case, we say that $(C[a,b],u)^* \cong BV[a,b]$, via the linear isometric isomorphism $L \longleftrightarrow g$ given by the integral representation:

$$L(f) = \int_a^b f dg \quad \text{for all } f \in C[a,b] \text{ and } \|L\| = V(g).$$

To state the Riesz representation Theorem for the dual of function space $C(S)$, S a topological space, we need the following terminology from "Measure Theory".

Definition: Let S be a given set. A collection $\mathcal{A} = \mathcal{A}(S)$ of subsets of S is called an **algebra** if it satisfies the following conditions:

(a) $S \in \mathcal{A}$;

(b) The union of finite number of sets in $\mathcal{A}(S)$ is also in $\mathcal{A}$;

(c) For any $A, B \in \mathcal{A}$, the difference set $A \backslash B$ is also in $\mathcal{A}$.

Clearly $\varnothing \in \mathcal{A}$. Further, $\mathcal{A}(S)$ is closed under finite intersections.

Definition: An algebra $\mathcal{A}(S)$ is called a σ-**algebra** if it closed under countable unions (hence also under countable intersections).

Definition: Let $\mathcal{A}(S)$ be an algebra of subsets of S. Then a function $\mu : \mathcal{A}(S) \to [-\infty, \infty] = \mathbb{R} \bigcup \{\pm\infty\}$ is called

(i) **finitely additive** if, for any $A_1, ..., A_n \in \mathcal{A}(S)$ with $A_i \cap A_j = \varnothing$ for $i \neq j$,

$$\mu(\bigcup_{i=1}^n A_i) = \sum_{i=1}^n \mu(A_i);$$

(ii) **countably additive** if, for any sequence of sets $A_1, A_2, ... \in \mathcal{A}(S)$ with $\bigcup_{i=1}^\infty A_i \in \mathcal{A}(S)$ and $A_i \cap A_j = \varnothing$ for $i \neq j$,

$$\mu(\bigcup_{i=1}^\infty A_i) = \sum_{i=1}^\infty \mu(A_i).$$

Definition: Let $\mathcal{A}(S)$ be a σ-**algebra** of subsets of S. A set function $\mu : \mathcal{A}(S) \to [0, \infty]$ is called a **positive measure** on S (or $\mathcal{A}(S)$) if

(i) $\mu(\varnothing) = 0$;

(ii) μ is countably additive.

Definition: A set function $\mu : \mathcal{A}(S) \to [-\infty, \infty]$ is called a **signed measure** on S if

(i) $\mu(\varnothing) = 0$;

(ii) μ is countably additive.

Note that, in this case μ can assume only one of the values $\pm\infty$. By the well-known **Hahn-Jordan decomposition**, every sighned measure μ on S can be expressed as $\mu = \mu^+ - \mu^-$, where μ^+, μ^- are positive measures on S.

Note. Alternatively, if μ is a sighned measure on $\mathcal{A}(S)$, we can define the **set functions μ^+ and μ^- on $\mathcal{A}(S)$** by

$$
\begin{aligned}
\mu^+(A) &= \sup\{\mu(B) : B \in \mathcal{A}(S), B \subseteq A\}, \\
\mu^-(A) &= -\inf\{\mu(B) : B \in \mathcal{A}(S), B \subseteq A\} \\
&= \sup\{-\mu(B) : B \in \mathcal{A}(S), B \subseteq A\}.
\end{aligned}
$$

Then both μ^+, μ^- are positive measures on S and $\mu = \mu^+ - \mu^-$.

Definition: A set function $\mu : \mathcal{A}(S) \to \mathbb{K}$ ($= \mathbb{R}$ or $\mathbb{C}$) is called a **complex-valued measure** (or simply, a **measure**) on S if

(i) $\mu(\varnothing) = 0$;

(ii) μ is countably additive.

Note that, in this case μ does not assume any infinite values $\pm\infty$. Further, we can define the real and imaginary parts $Re\mu$, $Im\,\mu$ of μ by

$$
(Re\mu)(A) = Re(\mu(A)) \quad \text{and} \quad (Im\mu)(A) = Im(\mu(A)), \quad A \in \mathcal{A}(S).
$$

Then $Re\mu$ and $Im\,\mu$ are sighned measures on S and $\mu = Re\mu + i\,Im\,\mu$.

In view of the above terminology, any (complex-valued) measure μ on S can be expressed as:

$$
\mu = (\mu_1 - u_2) + i(\mu_3 - \mu_4),
$$

where $\mu_1, \mu_2, \mu_3, \mu_4$ are positive measures on S.

Definition: Let μ be a measure on S. The **total variation** of μ on S is defined by

$$
|\mu|(A) = \sup\{\sum_{i=1}^{n} |\mu(B_i)| : B_i \in \mathcal{A}(S), B_i \cap B_j = \varnothing \text{ for } i \neq j \text{ and } A = \bigcup_{i=1}^{n} B_i\},
$$

$A \in \mathcal{A}(S)$.

Note. The total variation $|\mu|$ is a positive measure on $\mathcal{A}(S)$. It is the smallest positive measure such that:

$$
|\mu(A)| \leq |\mu|(A) \text{ for all } A \in \mathcal{A}(S);
$$

i.e., $|\mu|$ is the smallest of the positive measures ν on S such that:

$$
|\mu(A)| \leq \nu(A) \text{ for all } A \in \mathcal{A}(S).
$$

If μ is a sighned measure on $\mathcal{A}(S)$ and $\mu = \mu^+ - \mu^-$, then $|\mu| = \mu^+ + \mu^-$.

Definition: Let S be a completely regular Hausdorff space, and let $Bo(S)$ denote the smallest σ-algebra of subsets of S containing open (or equivalently, closed) subsets of S. Clearly, $Bo(S)$ is closed under countable unions and intersections.

(i) The members of $Bo(S))$ are called **Borel sets**.

(ii) Any measure $\mu : Bo(S) \to \mathbb{K}$ is called a **Borel measure** on S.

Definition: A Borel measure μ on S is called **regular** if, for each $A \in Bo(S)$,

$$\begin{aligned} |\mu|(A) &= \sup\{|\mu|(K) : K \subseteq A, K \text{ a compact subset of } S\} \\ &= \inf\{|\mu|(G) : A \subseteq G, G \text{ an open subset of } S\}; \end{aligned}$$

i.e., for each $A \in Bo(S)$ and $\varepsilon > 0$, there exist K compact and G open with $K \subseteq A \subseteq G$, such that

$$|\mu|(H) < \varepsilon \text{ for all } H \in Bo(S) \text{ with } H \subseteq G\backslash K.$$

Theorem A.6.2. *Let $M(S) = M(Bo(S))$ be the set of all bounded regular Borel measures on $Bo(S)$. Then $M(S)$ is a vector space under the operations:*

$$(\mu + \nu)(A) = \mu(A) + \nu(A), \quad (a.\mu)(A) = a.\mu(A),$$

where $\mu, \nu \in M(S)$, $a \in \mathbb{K}$ and $A \in Bo(S)$. Furtrher, $\|\mu\| = |\mu|(S)$ defines a norm for which $M(S)$ is a Banach space.

The space of all positive regular Borel measures on S is denoted by $M^+(S)$.

Definition: If $\mu \in M^+(S)$ and $f \in C_b(S, \mathbb{R})$. If $f \geq 0$, we define the integral of f w.r.t μ by

$$\int_S f d\mu = \sup\left\{ \int_S \varphi d\mu : 0 \leq h \leq f, \quad \varphi \text{ a simple function w.r.t. } Bo(S) \right\}.$$

For arbitrary $f \in C_b(S, \mathbb{R})$, we can write $f = f^+ - f^-$, where $f^+(x) = \max\{f(x), 0\}$ and $f^-(x) = \max\{-f(x), 0\}$. Then we define

$$\int_S f d\mu = \int_S f^+ d\mu - \int_S f^- d\mu.$$

Similarly, if $\mu \in M(S)$ and $f \in C_b(S) = C_b(S, \mathbb{C})$, we can define the integral $\int_S f d\mu$, as follows. We first need to write $\mu = (\mu_1 - \mu_2) + i(\mu_3 - \mu_4)$, where $\mu_1, \mu_2, \mu_3, \mu_4 \in M^+(S)$, and also $f = f_1 + if_2$, where $f_1, f_2 \in C_b(S, \mathbb{R})$; hence

$$\int_S f d\mu = \left(\int_S f_1 d\mu_1 - \int_S f_1 d\mu_2 \right) + i\left(\int_S f_2 d\mu_3 - \int_S f_2 d\mu_4 \right).$$

The following version of the classical Riesz representation theorem is well-known (see [Rud87, Roy88, DS58]).

Theorem A.6.3A (**Riesz-Markov, 1938**) *Let S be a compact Hausdorff space. Then $(C(S), u)^* \cong M(S)$ via the linear isometric isomorphism $L \longleftrightarrow \mu$, where*

$$L(f) = \int_S f d\mu \text{ for all } f \in C(S) \text{ and } ||L|| = |\mu|(S).$$

Theorem A.6.3B. (**Riesz-Alexandroff**) *Let S be a completely regular Hausdorff space. Then $(C_b(S), u)^* \cong M(\beta S)$ via the linear isometric isomorphism $L \longleftrightarrow \mu$, where*

$$L(f) = \int_S f d\mu \text{ for all } f \in C_b(S) \text{ and } ||L|| = |\mu|(S).$$

Proof. This follows from the fact that $(C_b(S), u)$ is isometrically isomorphic to $(C_b(\beta S), u)$; hence by above theorem, $(C_b(S), u)^* \cong (C_b(\beta S), u)^* \cong M(\beta S)$.

Theorem A.6.3C. (**Riesz-Kakutani, 1940**) *Let S be a locally compact Hausdorff space. Then $(C_0(S), u)^* \cong M(S)$ via the linear isometric isomorphism $L \longleftrightarrow \mu$, where*

$$L(f) = \int_S f d\mu \text{ for all } f \in C_0(S) \text{ and } ||L|| = |\mu|(S).$$

Proof. See, [**Rud87**], Theorem 6.19).

Theorem A.6.5. [**BDS, p. 97**] *Let $(S, \mathcal{A}(S), \mu)$ be a measurable space, and let $|\mu| : \mathcal{A}(S) \to \mathbb{R}$ be the total variation on $\mathcal{A}(S)$. Then for any $A \in A$,*

$$|\mu|(S) = \sup_{E \in A(S)} |\mu(E)|.$$

Definition. [**RF10, p. 342**] By a **signed measure** ν on the measurable space $(S, \mathcal{A}(S))$ we mean an extended real-valued set function $\nu : \mathcal{A}(S) \to [-\infty, +\infty]$ that possesses the following properties:

(i) ν assumes at most one of the values $+\infty$, $-\infty$.

(ii) $\nu(\varnothing) = 0$.

(iii) For any countable collection $\{E_n\}$ of disjoint measurable sets,

$$\nu\left(\bigcup_{n=1}^{\infty} E_n\right) = \sum_{n=1}^{\infty} \nu(E_n),$$

where the series $\sum_{n=1}^{\infty} \nu(E_n)$ converges absolutely if $\nu\left(\bigcup_{n=1}^{\infty} E_n\right)$ is finite.

Remark. A measure is a special case of a signed measure. It is not difficult to see that the difference of two measures, one of which is finite, is a signed measure. In fact, the forthcoming Jordan Decomposition Theorem will tell us that every signed measure is the difference of two such measures.

Definition. [RF10, p. 343] Let ν be a signed measure. We say that a set A is **positive** (with respect to ν) provided A is measurable and for every measurable subset E of A we have $\nu(E) > 0$. The restriction of ν to the measurable subsets of a positive set is a measure. Similarly, a set B is called **negative** (with respect to ν) provided it is measurable and every measurable subset of B has nonpositive ν measure. The restriction of $-\nu$ to the measurable subsets of a negative set also is a measure. A measurable set is called **null** with respect to ν provided every measurable subset of it has ν measure zero.

Remark [RF10, p. 343]. The reader should carefully note the distinction between a null set and a set of measure zero: While every null set must have measure zero, a set of measure zero may well be a union of two sets whose measures are not zero but are negatives of each other. By the monotonicity property of measures, a set is null with respect to a measure if and only if it has measure zero. Since a signed measure ν does not take the values ∞ and $-\infty$, for A and B measurable sets,

$$\text{if } A \subseteq B \text{ and } |\nu(B)| < \infty, \text{ then } |\nu(A)| < \infty.$$

Proposition A.6.4. [RF10, p. 343] Let ν be a signed measure on the measurable space $(S, \mathcal{A}(S))$. Then every measurable subset of a positive set is itself positive and the union of a countable collection of positive sets is positive.

Theorem A.6.5. (Hahn Decomposition of X) [RF10, p. 344] *Let v be a signed measure on the measurable space $(S, \mathcal{A}(S))$. Then there is a positive set A for v and a negative set B for v for which*

$$S = A \bigcup B \text{ and } A \cap B = \varnothing.$$

Definition. [RF10, p. 345] A decomposition of S into the union of two disjoint sets A and B for which A is positive for ν and B negative is called a **Hahn decomposition** for ν. The preceding theorem tells us of the existence of a Hahn decomposition for each signed measure. Such a decomposition may not be unique. Indeed, if $\{A, B\}$ is a Hahn decomposition for ν, then by excising from A a null set E and grafting this subset onto B we obtain another Hahn decomposition $(A \sim E, B \bigcup E)$.

If $\{A, B\}$ is a Hahn decomposition for ν, then we define two measures v^+ *and* v^- with $v = v^+ - v^-$ by setting

$$v^+(E) = \nu(E \cap A) \text{ and } v^-(E) = -\nu(E \cap A).$$

Two measures v_1 and v_2 on $(S, \mathcal{A}(S))$ are said to be **mutually singular** (in symbols $v_1 \perp v_2$) if there are disjoint measurable sets A and B with $S = A \bigcup B$ for which $v_1(A) = v_2(B) = 0$.

The measures v^+ *and* v^- defined above are mutually singular:

Theorem A.6.6. (Jordan Decomposition of ν) [RF10, p. 345] *Let v be a signed measure on the measurable space $(S, A(S))$. Then there* *are two mutually singular measures* v^+ *and* v^- *on* $(S, A(S))$ *such that* $v = v^+ - v^-$. *Moreover, there is only one such pair of mutually singular measures.*

Definition. [**RF11, p. 345**] The decomposition of a signed measure ν given by this theorem is called the *Jordan decomposition* of ν. The measures v^+ *and* v^- are called the **positive and negative parts** (or **variations**) of ν. Since ν assumes at most one of the values $+\infty$ and $-\infty$, either v^+ or v^- must be finite. If they are both finite, we call ν a *finite signed measure*. The measure $|\nu|$ is defined on $\mathcal{A}(S)$ by

$$|\nu|(E) = v^+(E) + v^-(E) \quad \text{for all } E \in \mathcal{A}(S).$$

Theorem A.6.7. (Squeeze Approximation by simple functions) [**RF10, p. 363**] *Let* $(S, A(S))$ *be a measurable space and* f *a measurable function on* S *that is bounded on* S. *Then for each* $\varepsilon > 0$, *there are simple functions* φ_ε *and* ψ_ε *defined on* S *that have the following approximation properties:*

$$\varphi_\varepsilon \leq f \leq \psi_\varepsilon \ \text{ and } \ 0 \leq \psi_\varepsilon - \varphi_\varepsilon < \varepsilon \ \text{ on } \ S.$$

Theorem A.6.8. (Approximation by simple functions) [**RF10, p. 363**] Let $(S, A(S), \mu)$ be a measure space and f a measurable function on S. Then there is a sequence $\{\varphi_n\}$ of simple functions on S that converges pointwise on S to f and has the property that

$$|\varphi_n| \leq |f| \text{ for all } n \geq 1.$$

(i) If S is σ–finite, then we may choose the sequence $\{\varphi_n\}$ so that each φ_n vanishes outside a set of finite measure.

(ii) If f is nonnegative, we may choose the sequence $\{\varphi_n)$ to be increasing and each $\varphi_n > 0$ on S.

A precise realization of the last of Littlewood's principle is the following surprising theorem.

Theorem A.6.9 (Egoroff) [**RF10, p. 364**] *Let* $(S, A(S), \mu)$ *be a measure space, and let* $E \in A(S)$ *be a set of finite measure. Let* $\{f_n\}$ *be a sequence of measurable functions on* E *that converges pointwise on* E *to the real-valued function* f *a.e. Then, for each* $\varepsilon > 0$, *there is a measurable set* $A \subseteq S$ *such that and* $f_n \to f$ *uniformly on* A *and* $\mu(E \backslash A) < \varepsilon$

In particular, we have

Theorem A.6.10 (Egoroff) [**RF10, p. 364**] *Let* $(S, A(S), \mu)$ *be a finite measure space. Let* $f_n\}$ *be a sequence of measurable functions on* S *that converges pointwise to the real-valued function* f *a.e. Then, for each* $\varepsilon > 0$, *there is a measurable set set* $A \subseteq S$ *such that and* $f_n \to f$ *uniformly on* A *and* $\mu(S \backslash A) < \varepsilon$

Theorem A.6.11. (Fatou Lemma) [**RF10, p. 368**] *Let* $\{f_n\}$ *be a sequence of nonnegative measurable functions on a measure space* $(S, \mathcal{A}(S), \mu)$

such that $f_n \to f$ a. e. on a measurable set $E \subseteq S$. Then

$$\int_E f.d\mu \leq \underline{\lim} \inf \int_E f_n d\mu.$$

Theorem A.6.12. (Monotone Convergence Theorem) [**RF10, p. 370**]
Let $\{f_n\}$ be a sequence of nonnegative measurable functions on a measure space $(S, \mathcal{A}(S), \mu)$ such that:
(a) $f_n \to f$ a.e;
(b) $f_n \leq f$ for all $n \geq 1$.
Then

$$\lim_{n \to \infty} \int_S f_n.d\mu = \int_S f.d\mu.$$

Corollary A.6.13. *Let $\{f_n\}$ be a sequence of nonnegative measurable functions on a measure space $(S, \mathcal{A}(S), \mu)$. Then*

$$\int_S \sum_{n=1}^{\infty} f_n d\mu = \sum_{n=1}^{\infty} \int_S f_n.d\mu.$$

Definition. [**RF10, p. 371**] Let $(S, \mathcal{A}(S), \mu)$ be a measure space and f a nonnegative measurable function on S. Then f is said be *integrable* over S with respect to μ provided $\int_S f.d\mu < \infty$.

Theorem A.6.14. (Dominated Convergence Theorem) [**RF10, p. 376**]
Let $\{f_n\}$ be a sequence of nonnegative measurable functions on a measure space $(S, \mathcal{A}(S), \mu)$ such that
(a) $f_n \to f$ a. e. on a measurable set $E \subseteq S$.
(b) There exists an integrable function $g \geq 0$ such that $|f_n(x)| \leq g(x)$ for all $x \in E$ and $n \geq 1$.
Then

$$\lim_{n \to \infty} \int_E f_n d\mu = \int_E f d\mu.$$

Theorem A.6.15 (Vitali Convergence Theorem). [RF10, p. 377] Let $(S, \mathcal{A}(S), \mu)$ be a measure space and $\{f_n\}$ a sequence of scaler functions on S that is both uniformly integrable and tight over S. Assume $f_n \to f$ pointwise a.e. on S and the function f is integrable over S. Then

$$\lim_{n \to \infty} \int_E f_n d\mu = \int_E f d\mu.$$

Definition. [**RF10, p. 381**] Let $(S, \mathcal{A})$ be a measurable space. For μ a measure on $(S, \mathcal{A})$ and f a nonnegative function on S that is measurable with respect to $\mathcal{A}$, define the **set function** $\nu : \mathcal{A}(S) \to \mathbb{R}^*$ on S by

$$\nu(E) = \int_E f.d\mu, \; E \in \mathcal{A}(S).$$

Then by the linearity of integration and the Monotone Convergence Theorem that ν is a measure on the measurable space $(S, \mathcal{A})$, and it has the property that

$$\text{if } E \in \mathcal{A}(S) \text{ and } \mu(E) = 0, \text{ then } \nu(E) = 0.$$

Definition. [RF10, p. 381] A measure ν is said to be **absolutely continuous** with respect to the measure μ provided (29) holds. We use the symbolism $\nu << \mu$ for ν absolutely continuous with respect to μ.

The following proposition recasts absolute continuity in the form of a familiar continuity criterion.

Proposition A.6.16. [RF10, p. 381] Let $(S, \mathcal{A}, \mu)$ be a measure space and ν a finite measure on the measurable space $(S, \mathcal{A})$. Then ν is absolutely continuous with respect to μ if and only if for each $\varepsilon > 0$, there is a $\delta > 0$ such that for any set $E \in \mathcal{A}(S)$,

$$\text{if } \mu(E) < \delta, \text{ then } \nu(E) < \varepsilon.$$

Theorem A.6.17 (Radon-Nikodym) **[RF10, p. 382]** *Let* $(S, A(S), \mu)$ *be a totally finite measure space and* $(S, \mathcal{A}(S), \mu)$ *a signed measure space. If* $\nu << \mu$, *there exists an integrable function* g *on* S *such that*

$$\nu(E) = \int_E g\, d\mu, \ E \in \mathcal{A}(S).$$

Definition. [RF10, p. 390] Let $(S, \mathcal{A})$ be a measurable space. A sequence $\{\nu_n\}\}$ of measures on $\mathcal{A}$ is said to **converge set-wise** on $\mathcal{A}$ to the set function ν provided

$$v(E) = \lim_{n \to \infty} \nu_n(E), \ E \in \mathcal{A}(S).$$

Definition. [RF10, p. 390] Let $(S, A(S), \mu)$ be a finite measure space. A sequence $\{\nu_n\}$ of finite measures on $\mathcal{A}$, each of which is absolutely continuous with respect to μ, is said to be **uniformly absolutely continuous** (or **equi absolutely continuous**) with respect to μ provided for each $\varepsilon > 0$, there is a $\delta > 0$ such that for any measurable set E and any $n \geq 1$,

$$\text{if } \mu(E) < \delta, \text{ then } \nu_n(E) < \varepsilon.$$

Theorem A.6.18 (Vitali-Hahn-Saks) **[RF10, p. 392]** Let $(S, \mathcal{A}(S), \mu)$ be a finite measure space and $\{\nu_n\}$ a sequence of finite measures on S, each of which is absolutely continuous with respect to μ. Suppose that $\{\nu_n(S)\}$ is bounded and $\{\nu_n\}$ converges setwise on S to ν. Then the sequence is uniformly absolutely continuous with respect to μ. Moreover, ν is a finite measure on S that is absolutely continuous with respect to μ.

Theorem A.6.19. (Nikodym) **[DS58, p. 160; RF10, p. 392]** Let $(S, \mathcal{A}(S))$ be a measurable space and $\{\nu_n\}$ a sequence of finite measures on

S that converges set-wise on S to the set function v. Assume $\{\nu_n(S)\}$ is bounded. Then ν is a measure on S and the σ-additivity of ν_n is uniform for $n = 1, 2, \ldots$

Definition. Let $(S, \mathcal{A}(S), \mu)$ be a finite measure space. For $0 < p < 1,$ the space

$$L_p(S, \mu) = \{f : S \longrightarrow \mathbb{K} : \int_X |f(x)|^p \, d\mu < \infty\},$$

the set of all Lebesgue p-integrable functions,. Then $L_p(S, \mu)$ is a Banach space with norm given by

$$||f||_p = \left(\int_X |f(x)|^p \, d\mu \right)^{1/p} , \quad f \in L_p(S, \mu)$$

Definition. [**RF10, p. 399**] For $1 < p < \infty$, let f belong to $L_q(S, \mu)$, where $q > 1$ *with* $\frac{1}{p} + \frac{1}{q} = 1$. Define the linear functional $T_f : L_p(S, \mu) \to \mathbb{R}$ by

$$T_f(g) = \int_S fg d\mu \quad \text{for all } g \in L_p.$$

Hölder's Inequality tells us that T_f is a bounded linear functional on L_p and its norm is at most $||f||_q$ while (1) tells us that its norm is actually equal to $||f||_q$ Therefore $T : L_q(S, \mu) \to (L_p(S, \mu))^*$ is an isometry. In the case that S is a Lebesgue measurable set of real numbers and μ is Lebesgue measure, we proved that T maps $L_q(S, \mu)$ onto $(L_p(S, \mu))^*$, that is, every bounded linear functional on $L_p(S, \mu)$ is given by integration against a function in $L_q(S, \mu)$. This fundamental result holds for general σ-finite measure spaces.

Theorem A.6.20 (Riesz Representation Theorem) [RF10, p. 400] (for Lp-spaces) *Let $(S, A(S), \mu)$ be a σ-finite measure space and $p, q > 1$ with $\frac{1}{p} + \frac{1}{q} = 1$. Then, $(L_p)^* = L_q$ via $T \longleftrightarrow g$, where $T \in (L_p)^*$ and $g \in L_q$ such that*

$$T(f) = \int_S fg d\mu \quad \text{for all } f \in L_p.$$

Appendix B

Bibliography

[AB98]   C.D. Aliprantis and O. Burkinshaw, Principles of real analysis (Third edition, Academic Press, 1998).

[Ap57]   T.M. Apostal, *Mathematical Analysis* (Addison-Wesley, 1957).

[BDS55]   R.G.Bartle, N. Dunford and J.T. Schwartz, Weak Compactness and Vector Measures, Canad. J. Math. 7(1955), , 289-305.

[BoD01]   B. Bongiorno and N. Dinculeanu, *The* Riesz Representation Theorem and Extension of Vector valued Additive Measures, J. Math. Anal. Appl. **261(2001)**, 706-732.

[Bou]   N. Bourbaki, Eléments de mathématique. XXV. Part I. Livre VI: Intégration. Chap. 6: Intégration vectorielle, Actualite's Sci. Indust., no. 1281, Hermann, Paris 19-59.

[BL74]   J.K. Brooks and P.L. Lewis, Linear Operators and Vector Measures, Trans. Amer. Math. Soc. 192(1974), 139-162.

[Buc58]   R.C. Buck, Bounded continuous functions on a locally compact space, Michigan Math. J. 5(1958), 95-104.

[Buc74]   R.C. Buck, Approximation properties of vector-valued functions, Pacific J. Math. 53(1974), 85-94.

[CDM61]   N. Colojoara, N. Dinculeanu and Gh. Marinescu, Measures with values in locally convex spaces, Bull. Math. Soc. Sei. Math. Phys. R. P. Roumaine 5 (53) (1961) 167-180.

[DU77]   J. Diestel, J.J. Uhl, *Vector Measures,* Math. Surveys No 15, AMS,1977.

[Din67]   N. Dinculeanu, *Vector Measures,* Pergamon Press.1967.

[Dob71]   I. Dobrakov, On Representation of Linear Operators on $C_0(S, X)$, Czech. Math. J. 21(1971), 13-30.

[DL13]   L. Drewnowskia and I. Labudab, Bartle–Dunford–Schwartz integration, J. Math. Anal. Appl. 401(2013) 620–640.

[DS58]   N. Dunford and J.T. Schwartz, *Linear operators, Part I: General Theory* (Interscience, 1958).

[DRS69]   P. L. Duren, B. W. Romberg and A. L. Shields, Linear functionals on Hp spaces with $0 < p < 1$, J. Reine Angew. Math. 238 (1969), 32-60.

[EW71] J. R. Edwards and S. G. Wayment, Integral representations for continuous linear operators in the setting of convex topological vector spaces, Trans. Amer. Math. Soc. 157 (1971), 329-346.

[Edw65] R. E. Edwards, *Functional Analysis, Theory and Applications* (Holt, Rinehart and Winston, 1965).

[FS60] C. Foias and I. Singer, Some remarks on the representation of linear operators in spaces of vector-valued continuous functions, Rev. Math. Pures Appl. 5 (I960), 729-752.

[Fol99] G.B. Folland, Real Analysis: Modern Techniques and their Applications, 2nd ed. Wiley, New York, 1999.

[Fon74] R.A. Fonetnot, Strict topologies for vector-valued functions, Canad. J. Math. 26(1974), 841-853; Corrigendum, Canad. J. Math. 27(1975), 1183-1184.

[Gar73] D. Garling, A short proof of the Riesz representation theorem, Proc. Camb. Plil. Soc. 73(1973), 459-460.

[Gil71] R. Giles, A generalization of the strict topology, Trans. Amer. Math. Soc. 161(1971), 467-474.

[GJ60] L. Gillman and M. Jerison, *Rings of Continuous Functions* (D.Van Nostrand, 1960).

[Goo70] R.K. Goodrich, A Riesz Representation Theorem, Proc. Amer. Math. Soc. 24(1970), 629-636.

[Gra66] B. Gramsch, Integration und holomorphe Funktionen in lokalbeschrankten Raumen, Math. Ann. 162 (1965/66), 190-210.

[Gra67] B. Gramsch, Tensorprodukte und Integration vektorwertiger Funktionen, Math. Z. 100 (1967), 106-122.

[GreS70] D. A. Gregory and J. H. Shapiro, Nonconvex linear topologies with the Hahn-Banach extension property, Proc. Amer. Math. Soc. 25 (1970), 902-905.

[Gro53] A. Grothendieck, Sur les applications linéaires faiblement compactes d'espaces du type $C(K)$. (French) Canadian J. Math. 5, (1953). 129-173.

[Gu72] D. Gulick, The σ-compact-open topology and its relatives, Math. Scand. 30(1972), 159-176.

[Hal50] P.R. Halmos, *Measure Theory* (D. Van Nostrand, 1950).

[Hen96] W.A. Hensgen, A simple proof of Singer's Represemtation Theorem, Proc. Amer. Math. Soc. 124(10)(1996), 3211-3212.

[HP57] E. Hille and R.S. Phillips. Functional Analysis and Semigroups (AMS Colloquium, 1957).

[Hor66] J. Horvath, *Topological Vector Spaces and Distributions* (Addison-Wesley, 1966).

[Kak41] S. Kakutani, Concrete representation of abstract (M)-spaces, Ann. Math. 42(1941), 994-1024.

[Kat74] A.K. Katsaras, Continuous linear functionals on spaces of vector-valued functions, Bull. Soc. Math. Grece 15(1974), 13-19.

[Kat75] A.K. Katsaras, Spaces of vector measures, Trans. Amer. Math. Soc. 206(1975), 313-328.

[Kat76] A.K. Katsaras, Locally convex topologies on spaces of continuous vector functions, Math. Nachr. 71(1976), 211-226.

[Kel55] J.L. Kelley, *General Topology* (Van Nostrand, 1955).

[KN76] J.L. Kelley, I. Namioka and Co-authors, *Linear Topological Spaces*, (D. Van Nostrand, 1963; Springer-Verlag, 1976).

[Kha79] L.A. Khan, The strict topology on a spaces of vector-valued functions, Proc. Edinburgh Math. Soc. 22(1979), 35-41.

[Kha93] L.A. Khan, Integration of vector-valued continuous functions and the Riesz representation theorems, Studia Sci. Math. Hungarica 28(1993), 71-77.

[Kha95] L.A. Khan, Some approximation results for the compact-open topology, Periodica Math. Hungarica 30(1995), 81-86.

[KR81] L.A. Khan and K. Rowlands, On the representation of strictly continuous linear functionals, Proc. Edinburgh Math. Soc. 24(1981), 123-130.

[KR91] L.A. Khan and K. Rowlands, The σ-compact-open topology and its relatives on a space of vector-valued functions, Boll. Unione Mat. Ital. (7)5-B (1991), 729-739.

[Khu78a] S.S. Khurana, Topologies on spaces of vector-valued continuous functions, Trans. Amer. Math. Soc. 241(1978), 195-211.

[Khu78b] S.S. Khurana, Topologies on spaces of vector-valued continuous functions II, Math. Ann. 234(1978), 159-166.

[Khu09] S.S. Khurana, Integral Representation for a class of Operators,J. Math. Anal. Appl. 350(2009), 290-293.

[Klu67] I. Kluvanek, Characterization of Fourier-Stieltjes Transformsof vector measures, Czech. Math. J. 17(1967), 261-277

[Ko69] G.Köthe, *Topological Vector Spaces I* (Springer-Verlag, 1969).

[MO53] S. Mazur and W. Orlicz, Sur les espaces métriques linéaires. I, II, Studia Math. 10(1948), 184-208; ibid. 13 (1953), 137-179. MR 10, 611.

[Mez02] L. Meziani, Integral representation for a class of vector valued operators, Proc. Amer. Math. Soc. 130(2002). 2067-2077.

[Mez05] L. Meziani, Some change of variable formulas in integral representation theory. Acta Math. Univ. Comenian. (N.S.) 74 (2005), no. 1, 59–70.

[Mez08] L. Meziani, A theorem of Riesz type with Pettis integrals in topological vector spaces, J. Math. Anal. Appl. 340(2008), 817-824.

[Mez09a] L. Meziani, A simple integral representation for bounded operators in topological vector spaces. Appl. Math. Sci. (Ruse) 3 (2009), no. 17-20, 861–868.

[Mez09b] L. Meziani, Tightness of probability measures on function spaces. J. Math. Anal. Appl. 354 (2009), no. 1, 202–206.

[MAW08] L. Meziani, S. Almezel and M.N. Waly, The integral structure of some bounded operators on $L_1(\mu, X)$. Int. J. Math. Anal. (Ruse) 2 (2008), no. 9-12, 437–446.

[MWA09] L. Meziani, M.N. Waly, and S. Almezel, Characterizing linear bounded operators via integral. Int. Math. Forum 4 (2009), no. 5-8, 233–240.

[Naim72] M. A. Nairmark, Normed Algebras: Wolters-Noordhoff, 1972.

[Pie65] A. Pietsch, Nukleare lokalkonvexe Raume, Schriftenreihe der Institute für Mathematik bei der Deutschen Akademie der Wissenschaften zu Berlin, Reihe A, Reine Mathematik, Heft 1, Akademie-Verlag, Berlin, 1965.

[Po66] V. Popescu, Integration with respect to measures with values in arbitrary topological vector spaces, Stud. Cetc. Mat. 18 (1966), 1159-1180.

[Pro77] J.B. Prolla, *Approximation of Vector-valued Functions* (North-Holland Math. Studies No. 25, 1977).

[PRR66] D. Przeworska-Rolewicz and S. Rolewicz, Ozz integrals of functions with values in a complete linear metric space, Studia Math. 26 (1966), 121-131.

[Rie09] F.Riesz, Sur les opérations fonctionnelles linéaires, C.R.Acad. Sci. Paris, 149(1909), 974-977.

[RR64] A.P. Robertson and W. Robertson, *Topological Vector Spaces* (Cambridge University Press, 1964).

[Rol85] S. Rolewicz, *Metric Linear Spaces*, (D. Reidel Publishing Company, 1985).

[Roy88] H.L. Royden, *Real Analysis* (Third Edition, MacMillan, 1988).

[RF10] H. L. Royden and P. M. Fitzpatrick, *Real Analysis* (Fourth Edition, Macmillan, 2010).

[Rud87] W. Rudin, *Real and Complex Analysis* (McGraw-Hill, 1966; Third Edition,1987).

[Rud91] W. Rudin, *Functional Analysis* (McGraw Hill, 1973; Second Revised Edition,1991).

[Sch71] H.H. Schaefer, *Topological Vector Spaces* (Macmillan, 1966; Springer-Verlag, 1971).

[Sen72] F.D. Sentilles, Bounded continuous functions on a completely regular space, Trans. Amer. Math. Soc. 168(1972), 311-336.

[Shu69] A. H. Shuchat, Integral representation theorems in topological vector spaces, Dissertation, University of Michigan, Ann Arbor, Mich., 1969.

[Shu72a] A.H. Shuchat, Approximation of vector-valued continuous functions, Proc. Amer. Math. Soc., 31(1972), 97-103.

[Shu72b] A.H. Shuchat, Integral representation theorems in topological vector spaces, Trans. Amer. Math. Soc. 172(1972), 373-397.

[Sing57] I.Singer, Linear Functionals on the space of continuous mappings of a compact space into Banach space, Rev.Math.Pures.Appl ,2(1957) ,301-315 (Russian).

[Ste65] E. F. Steiner, On finite-dimensional linear topological spaces, Amer. Math. Monthly 72 (1965), 34-35.

[Swa73] C. Swartz, Absolutely summing and dominated operators on spaces of vector-valued continuous functions, Trans. Amer. Math. Soc. 179 (1973), 123–131.

[Swo64] K. Swong, A representation theory of continuous linear maps, Math. Ann. 155 (1964), 270-291; errata, ibid. 157 (1964), 178.

[Thom12] E.G.F. Thomas, Vector integration, Quaestiones Mathematicae, 35(4)(2012), 391-416.

[Thor68] B. L. D. Thorp, Equivalent notions of bounded variation, J. London Math. Soc. 43 (1968), 247-252.

[TW68] P. Turpin and L. Waelbroeck, Szzr l'approximation des fonctions difrenliables a valeurs dans les espaces vectoriels topologiques, C. R. Acad. Sei. Paris Ser. A-B 267 (1968), A94-A97.

[Ula68] M. P. Ulanov, Vector-valued set functions and the representation of continuous linear mappings, Sibirsk. Mat. Z. 9 (1968), 410-425 (Russian).

[Var65] V.S. Varadarajan, Measures on topological spaces, Mat. Sb(N.S.) 55(97) (1961), 35-100; English transl. Amer. Math. Soc. Transl. (2) 48(1965), 161-228.

[Vog67] D. Vogt, Integrationstheorie in p-normierten Räumen, Math. Ann. 173 (1967), 219-232.

[Wel65] J. Wells, Bounded continuous vector-valued functions on a locally compact space, Michigan Math. J. 12(1965), 119-126.

[Wh83] R.F. Wheeler, A survey of Baire measures and strict topologies, Expositiones Math. 2(1983), 97-190.

[Wila78] A. Wilansky, *Modern Methods in Topological Vector Spaces* (McGraw-Hill, 1978).

[Will70] S.Willard, *General Topology* (Addison-Wesely, 1970).

[Won92] Y-C. Wong, *Introductory Theory of Topological Vector Spaces* (Marcel Dekker, 1992).

Appendix C

List of Symbols

$\overline{A}$ or $cl(A)$, the closure of a subset $A \subseteq S$,

A^i or $int\ A$, the interior of a subset $A \subseteq S$,

A^0, polar of a subset $A \subseteq X$,

$\mathcal{B}a(S)$, the σ-algebra of all Baire subsets of S,

$\mathcal{B}o(S)$, the σ-algebra of all Borel subsets of S,

$b = b(X, X^*)$, strong topology on X,

$b^* = b^*(X^*, X)$, strong topology on X^*,

β, the strict topology on $C_b(S), C_b(S, X)$,

βS, the Stone-Cech compactification of S,

$\mathbb{C}$, the field of complex numbers,

$C(S)$, the algebra of all continuous functions $\varphi : S \to \mathbb{K}$,

$C_b(S) = \{\varphi \in C(S) : \varphi \text{ is bounded on } S\}$,

$C_0(S) = \{\varphi \in C(S) : \varphi \text{ vanishes at infinity}\}$,

$C_{00}(S) = \{\varphi \in C(S) : \varphi \text{ has compact support}\}$,

$C(S, X)$, the vector space of all continuous functions $f : S \to X$,

$C_b(S, X) = \{f \in C(S, X) : f \text{ is bounded on } S\}$,

$C_0(S, X) = \{f \in C(S, X) : f \text{ vanishes at infinity}\}$,

$C_{00}(S, X) = \{f \in C(S, X) : f \text{ has compact support}\}$,

$C_{rc}(S, X) = \{f \in C(S, X) : f(S) \text{ is relatively compact in } X\}$,

$C(S) \otimes X = \{\varphi \otimes a : \varphi \in C(S), a \in X\}$, where $(\varphi \otimes a)(x) = \varphi(x)a, x \in S$,

C_{XX}, a class of operators $T : C(S, X) \to X$,

F_σ, countable union of closed sets

$\|f\| = \sup_{x \in S} \|f(x)\|$, $f \in C_b(S, X)$ and $\|.\|$ a norm on X.

$\|f\|_p = \|p \circ f\| = \sup_{x \in S} p(f(x))$, $f \in C_b(S, X)$ and $p \in cs(X)$,

$$\|f\|_p = \left(\int_X |f(x)|^p \, d\mu \right)^{1/p}, \quad f \in L_p(S, \mu), p \geq 1,$$

G_δ, countable intersection of open sets,

$\int_x \varphi d\mu$, the integral of $\varphi \in C_t(S)$ with respect to a scalar measure $\mu : S \to \mathbb{K}$,

$\int_x dmf$, the integral of $f \in C_o(S, X)$ with respect to a vector measure $m : S \to X^*$,

k, the compact-open topology on $C(S, X)$ or $C_b(S, X)$,

$\mathbb{K}$, the field of scalars, always either the field $\mathbb{R}$ or the complex field $\mathbb{C}$,

$M_{r\sigma bv}(\mathcal{B}(S), X) = M_{r\sigma bv}(\mathcal{B}(S))$, 1.4

$M(\mathcal{B}a(S))$, the vector space of all bounded $\mathbb{R}$-valued σ-additive measures on $\mathcal{B}a(S)$,

$M_t(\mathcal{B}o(S))$, the vector space of all bounded $\mathbb{R}$-valued tight measures on $\mathcal{B}o(S)$,

$M(\mathcal{B}a(S), X^*)$, the vector space of all bounded X^*-valued σ-additive measures on $\mathcal{B}a(S)$,

$M_t(\mathcal{B}o(S), X^*)$, the vector space of all bounded X^*-valued tight measures on $\mathcal{B}o(S)$,

ρ_A, the Minkowski functional of $A \subseteq X$ defined by $\rho_A(x) = \inf\{\lambda > 0 : x \in \lambda A\}$,

$\varphi \otimes a$, the function $\varphi \otimes a : S \to X$, $(\varphi \otimes a)(x) = \varphi(x)a, \varphi \in C(S), a \in X$,

$\prod_{\alpha \in I} S_\alpha$, the Cartesian product of a family $\{S_\alpha : \alpha \in I\}$ of topological spaces with the product topology,

u, the uniform topology on $C_b(S, X)$,

$w(X, X^*)$ or $w(X, X^*)$, the weak topology on a TVS X,

$w(X^*, X)$ or $w(X^*, X)$, the weak*-topology on the dual space X^*,

X', the algebraic dual of a vector space X,

X^*, the topological (or continuous) dual of a topological vector space X,

Appendix D

Index

Printed by Books on Demand GmbH, Norderstedt / Germany